Advanced Temperature Measurement and Control

Second Edition

Advanced Temperature Measurement and Control
Second Edition

Gregory K. McMillan

ISA™

Notice

The information presented in this publication is for the general education of the reader. Because neither the author(s) nor the publisher has any control over the use of the information by the reader, both the author(s) and the publisher disclaim any and all liability of any kind arising out of such use. The reader is expected to exercise sound professional judgment in using any of the information presented in a particular application.

Additionally, neither the author(s) nor the publisher has investigated or considered the effect of any patents on the ability of the reader to use any of the information in a particular application. The reader is responsible for reviewing any possible patents that may affect any particular use of the information presented.

Any references to commercial products in the work are cited as examples only. Neither the author(s) nor the publisher endorses any referenced commercial product. Any trademarks or tradenames referenced belong to the respective owner of the mark or name. Neither the author(s) nor the publisher makes any representation regarding the availability of any referenced commercial product at any time. The manufacturer's instructions on use of any commercial product must be followed at all times, even if in conflict with the information in this publication.

International Society of Automation
67 Alexander Drive
P.O. Box 12277
Research Triangle Park, NC 27709

Printed in the United States of America.
10 9 8 7 6 5 4 3 2

ISBN 978-1-936007-38-7

Library of Congress Cataloging-in-Publication Data in process

Contents

Preface

Temperature is one of the four most common types of loops. While the other common loops (flow, level, pressure) occur more often, temperature loops are generally more difficult and important. It is the single most frequently stated type of loop of interest to users, and the concern for better control extends to the widest variety of industries.

Temperature is a critical condition for reaction, fermentation, combustion, drying, calcination, crystallization, extrusion, or degradation rate and is an inference of a column tray concentration in the process industries.

Tight temperature control translates to lower defects and greater yields during seeding, crystal pulling, and rapid thermal processing of silicon wafers for the semiconductor industry.

For boilers, temperature is important for water and air preheat, fuel oil viscosity, and steam superheat control. For incinerators, an optimum temperature often exists in terms of ensured destruction of hazardous compounds and minimum energy cost. For heat transfer fluids such as cooling tower, chilled water, brine, or Therminol™, good temperature control minimizes upsets to the users.

Temperature control in cold rooms is used to reduce the contamination and degradation rate in pharmaceutical, biochemical, beverage, and food research and production.

Temperature control in plant growth chambers is important for studying the effects of hybridization, genetic engineering, and plant growth regulators.

The great impact of temperature control on conductivity and pH loops was not widely realized until recently. For aqueous solutions of acids, bases, and salts, the inferred concentration from conductivity and the pH of the solution often changes at the rate of about twenty percent and four tenths of a pH unit, respectively, per ten degrees centigrade. These changes, which reflect a fundamental alteration in the solution due to different ion dissociation and mobility, should not be confused with temperature effects on the signal developed by the electrode, which are corrected for by conventional temperature compensators. Thus, temperature control can be important for raw, cooling tower, deaerator, boiler, and wastewater treatment.

Good temperature control is important during the research, reaction, separation, processing, and storage of products and feeds and is thus a key to product quality. It is also of importance for environmental control and energy conservation.

Tight temperature control can extend the life of process equipment (e.g., reactor glass lining, scrubber fiberglass trays, or furnace firebrick) by prevention of excursions beyond the temperature rating. Abrupt changes in coolant or steam flow can shock equipment and upset other utility users. Thus, it is also important to monitor the controller output and use methods (e.g., set point velocity limits and split-range, criss-cross prevention logic) to prevent rapid changes or oscillations. Since temperature control is typically achieved by the direct or indirect manipulation of heat flow into or out of the system, a reduction in the overshoot and oscillation of temperature loops can also correspond to a decrease in energy consumption.

Curiously, the slowness of the response of the temperature process is the biggest source of problems and opportunities for tight temperature control. The slowness makes it difficult to tune the controller because the persistence and patience required to obtain a good open- or closed-loop test exceeds the capability of most humans. At the same time, this slowness, in terms of a large major process time constant, enables gain settings larger than those permissible in other types of loop except for level.

The nonlinearity of the process further aggravates the tuning problem. The dependence of the process gain on operating conditions and load has been discussed but not simplified and quantified in enough detail to facilitate online compensation.

The slowness of the response of the thermocouple or resistance temperature detector (RTD) in a thermowell slows down the ability of the controller to identify and react to upsets and considerably affects all of the tuning settings (i.e., gain, reset, and rate) of many temperature loops.

Once a properly implemented temperature loop is correctly tuned, the control error is often less than the tolerance (error limits) of the sensor. If one considers that the accumulated error of an installed thermocouple or RTD system can be about five times larger than the error limits of the sensor, one realizes that system measurement error seriously limits temperature loop performance.

The major contributors to the measurement system error for thermocouples and RTDs will be detailed and new sensor technologies will be identified in Chapter 2. The user can in many cases reduce the error significantly by modification of the installation or by the use of a breakthrough in technology.

The temperature control error can be less than the system measurement error for large back mixed volumes (e.g., vessels and columns) if the controller gain is maximized to take advantage of the large process time constant. For volumes without appreciable back mixing (e.g., desuperheaters, exchangers, and extruders), the process dead time exceeds the process time constant. The controller gain is smaller and consequentially the control errors are larger for these dead time dominant loops.

1
Temperature Measurement

Learning Objectives

A. **Appreciate the importance of temperature measurements.**

B. **Understand the relative performance of thermocouples and resistance temperature detectors.**

C. **Gain an overview of optical pyrometer performance.**

D. **Learn about the latest technological advances in transmitters and sensors.**

E. **Find out how to select specific sensor and thermowell designs.**

F. **Understand important aspects of extension wire effects.**

G. **Gain some insight into best transmitter and communication options.**

H. **Learn the basic thermowell location and installation requirements.**

I. **Learn terminology and issues to intelligently discuss industrial applications.**

1-1. Introduction and Overview

Temperature is a measure of a material's internal molecular activity. As the level of molecular activity rises, the temperature of the material increases. *Hot* and *cold* are subjective, qualitative descriptions of a change in molecular activity.

Temperature is often the most important of the common measurements because it is an indicator of process stream composition and product quality. It would be nice if we had online analyzers throughout the process but the fact of the matter is that most plants have infrequent, offline lab analysis at best. In the chemical industry, nearly all loops that are controlling the composition of a unit operation use temperature as the primary controlled variable. Even when online or at-line analyzers exist, these are usually relegated to monitoring and the manual trim or optimization of temperature set points by supervisory or model predictive control. Temperature measurements are also essential for equipment protection and performance monitoring. Some examples of common unit operations that have a critical dependence upon tight (minimum variability) temperature control are:

- Bioreactors and fermentors
- Calciners and kilns
- Chemical reactors
- Columns (e.g., absorber, distillation, stripper)
- Condensers
- Crystallizers
- Dryers
- Evaporators
- Extruders
- Furnaces
- Superheaters and desuperheaters
- Vaporizers

Over the years, the need in the process industry for more consistent and accurate ways to describe temperature led to the invention of temperature-measuring devices, or *sensors*. Sensors use standard, universally recognized temperature scales. Because these scales rely on fixed points in nature (e.g., freezing point of water), they provide a way to describe temperature that is both objective and quantitative. The four temperature scales in use today are Fahrenheit, Celsius (also called Centigrade), Kelvin, and Rankine. In commercial applications, Fahrenheit and Celsius are the most commonly used scales.

In industrial environments, high process temperatures, pressures, and vibration make it necessary to have a robust temperature sensor. Fast response time, accuracy, and stability are also needed. While several types of temperature sensors are available, such as thermistors, infrared pyrometers, fiber optic, and others, the two most commonly used in the process measurement industry are resistance temperature detectors (RTDs) and thermocouples (TCs).

Comparison of Thermocouples and Resistance Temperature Detectors

In the process industry as a whole, 99% or more of the temperature loops use thermocouples (TCs) or resistance temperature detectors (RTD). The RTD provides sensitivity (minimum detectable change in temperature), repeatability, and drift that are an order of magnitude better than the thermocouple, as shown in Table 1-1 [1, 2]. Sensitivity and repeatability are 2 of the 3 most important components of accuracy. The other most important component, resolution, is set by the transmitter. Drift is important for extending the time between calibrations. The data in this table dates back to the 1970s and consequently doesn't include the improvements made in thermocouple sensing element technology and premium versus standard grades. However, the differences are so dramatic that the message is still the same.

A Resistance Temperature Detector (RTD) has a much better sensitivity and repeatability, a lower and more predictable drift, and a higher signal level than a thermocouple (TC).

Table 1-1 includes data on thermistors, which have seen limited use in the process industry despite their extreme sensitivity and fast (millisecond) response, primarily because of their lack of chemical and electrical stability. Thermistors are also highly nonlinear but this can be addressed by smart instrumentation.

For bare sensing elements, thermistors are much faster-responding than thermocouples, which are slightly faster than RTDs. This point rarely comes into play because for most industrial processes a 1 or 2 second additional lag time in a temperature loop is well within the uncertainty of the loop's dynamics. The secondary process time lags can easily change by 10 to 20 seconds for slight changes in operating conditions. Also, once these sensing elements are put inside a thermowell or protection tube (a closed-end metal tube that encapsulates and protects a temperature sensor from process flow, pressure, vibration, and corrosion), the fit, fill, material, and construction of the thermowell have the biggest impact on temperature measurement time lags, as noted in Tables 1-2a and 1-2b [1][3, 4]. Protection tubes like thermowells provide isolation of the element from the process but unlike thermowells, protection tubes do not necessarily provide a pressure tight attachment to a vessel, a tapered or stepped wall, or a tight fit of the element. Protection tubes may be ceramic for high temperature applications. The measurement lags from protection tubes are generally larger than for thermowells.

Table 1-1. Accuracy, range, and size of temperature sensing elements [1, 2]

Criteria	Thermocouple	Platinum RTD	Thermistor
Repeatability (°C)	1 - 8	0.02 - 0.5	0.1 - 1
Drift (°C)	1 - 20	0.01 - 0.1	0.01 - 0.1
Sensitivity (°C)	0.05	0.001	0.0001
Temperature Range (°C)	–200 - 2000	–200 - 850	–100 - 300
Signal Output (volts)	0 - 0.06	1 - 6	1 - 3
Power (watts at 100 ohm)	1.6×10^{-7}	4×10^{-2}	8×10^{-1}
Minimum Diameter (mm)	0.4	2	0.4

Table 1-2a. Dynamics of bare sensing elements [1][3, 4]

Bare Sensing Element Type	Time Constant (seconds)
Thermocouple 1/8-inch sheathed and grounded	0.3
Thermocouple 1/4-inch sheathed and grounded	1.7
Thermocouple 1/4-inch sheathed and insulated	4.5
Single Element RTD 1/8 inch	1.2
Single Element RTD 1/4 inch	5.5
Dual Element RTD 1/4 inch	8.0

The thermowell and protection tube design and material, the process heat transfer coefficient, and the fit of the sensor determine the temperature measurement speed of response more than the sensor type.

Table 1-2b. Dynamics of thermowells [1][3, 4]

Process Fluid Type	Fluid Velocity (feet per second)	Annular Clearance (inches)	Annular Fill Type	Time Constants (seconds)
Gas	5	0.04	Air	107 and 49
Gas	50	0.04	Air	93 and 14
Gas	150	0.04	Air	92 and 8
Gas	150	0.04	Oil	22 and 7
Gas	150	0.02	Air	52 and 9
Gas	150	0.005	Air	17 and 8
Liquid	0.01	0.01	Air	62 and 17
Liquid	0.1	0.01	Air	32 and 10
Liquid	1	0.01	Air	26 and 4
Liquid	10	0.01	Air	25 and 2
Liquid	10	0.01	Oil	7 and 2
Liquid	10	0.055	Air	228 and 1
Liquid	10	0.005	Air	4 and 1

There are many stated advantages for thermocouples, but if you examine them more closely you realize they are not as important as perceived for industrial processes. Thermocouples are more rugged than RTDs. However, the use of good thermowell or protection tube design and installation methods makes an RTD sturdy enough for even high-velocity stream and nuclear applications [5, 6]. Thermocouples appear to be less expensive until you start to include the cost of extension lead wire and the cost of additional process variability from less sensor sensitivity and repeatability.

Thermocouple extension wire and the consequences of drift in terms of process offsets and the need for more frequent calibration make the life cycle costs of a thermocouple (TC) larger than a Resistance Temperature Detector (RTD).

The minimum size of a thermocouple is much smaller. While a tiny sensor size is important for biomedical applications, miniature sensors are rarely useful for industrial processes.

The main reason to go to a thermocouple is if the temperature range is beyond what is reasonable for an RTD or you don't need the accuracy of an RTD. Thus, for temperatures above 850°C (1500°F), the clear choice is a thermocouple for a contacting temperature measurement. For temperatures within the range of the RTD the decision often comes down to whether the temperature is used for process control or just the monitoring of trends. If you have lots of temperatures for trending where errors of several degrees are unimportant, you could save money by going to thermocouples with transmitters mounted on the thermowell (integral mount) or nearby. If you are using temperature for process control, data analytics, statistical or neural network predictions, process modeling, or in safety systems, a properly protected and installed RTD is frequently the best choice for temperatures lower than 500°C (900°F). At temperatures above 500°C, changes in sensor sheath insulation resistance have caused errors of 10°C or more.

Optical Pyrometers

While thermocouples can be used at high temperatures in furnaces and wireless transmitters have expanded their use in rotating equipment, such as kilns and calciners, there is still a significant niche market for optical pyrometers in the process industry.

Non-contacting temperature measurements are needed for solid products that are moving, such as plastic webs, fiber spin lines, and paper sheets. Non-contacting temperature measurements are also useful for furnaces where the installation integrity and reliability of thermocouples is questionable.

Optical pyrometers are required when contact with the process is not possible or extreme process temperatures and conditions cause chemical attack, physical damage, or an excessive decalibration, dynamic, velocity, or radiation error of a TC or RTD (see Chapter 2).

While the market is small for optical pyrometers, an overview of the technology is merited to better understand some of the limitations and difficulties not commonly discussed in product catalogs. Equations are used to clarify the assumptions and quantify the effects of a non-ideal target, installation, intervening space, and viewing window.

For details on selection, configuration, installation, and calibration, the user should use information and application specialists from the manufacturer.

Pyrometers infer temperature from the optical radiation intensity at one or more wavelengths from a target. The most prevalent non-contacting temperature measurements devices in industry are single-color and ratio (2-color) pyrometers [1][7, 8, 9, 10]. The concept of blackbodies, graybodies, and non-graybodies is necessary to understand how the radiation seen by the pyrometer for a particular temperature varies with target and process conditions.

Blackbodies

A blackbody is a theoretical surface that absorbs all energy (does not reflect or transmit energy). A blackbody is also a perfect radiating body. The total net energy radiated by a blackbody is proportional to the difference between the target temperature (T_t) and the sensor temperature (T_s) each raised to the 4th power as detailed in Equation 1-1a (Stephan-Boltzmann Equation). The total energy radiated is the area under the curve in Figure 1-1 at the given temperature. The parameter K takes into account the areas of the target and sensor and the distance between them. For high temperatures, the sensor temperature is considered negligible so this temperature is dropped in Equation 1-1b but the K factor remains to take into account the characteristics of the sensor [1][7, 8, 9, 10].

Emissivity and Emittance

Process temperature targets do not behave as blackbodies. In Equation 1-1b, the total emissivity term is introduced as a multiplier to take into account the non-ideal radiating behavior of materials (e.g., aluminum, carbon, Monel, steel, titanium, zirconium, plastic, paper). The total emissivity is the ratio of the total energy radiated by a material to the total energy radiated by a blackbody. The total emissivities of most materials encountered in process applications at various temperatures can be found in tables in books with chapters on pyrometers.

The emissivity is a physical property of the material and is specified for a sample with a highly polished surface. The targets in industrial equipment (e.g., furnaces, calciners, and kilns) have different roughness, geometry, and sizes and are often covered with oxides and corrosion products. To take into account the effects of installation and operating conditions, a factor called *emittance* is used instead of emissivity. The total emittance is the ratio of total radiation for the physical and operating conditions of the target to the total radiation from a blackbody at a given temperature. In Equation 1-1c total emissivity is replaced with total emittance. Tables are available to give some typical values of total emittance but the actual emit-

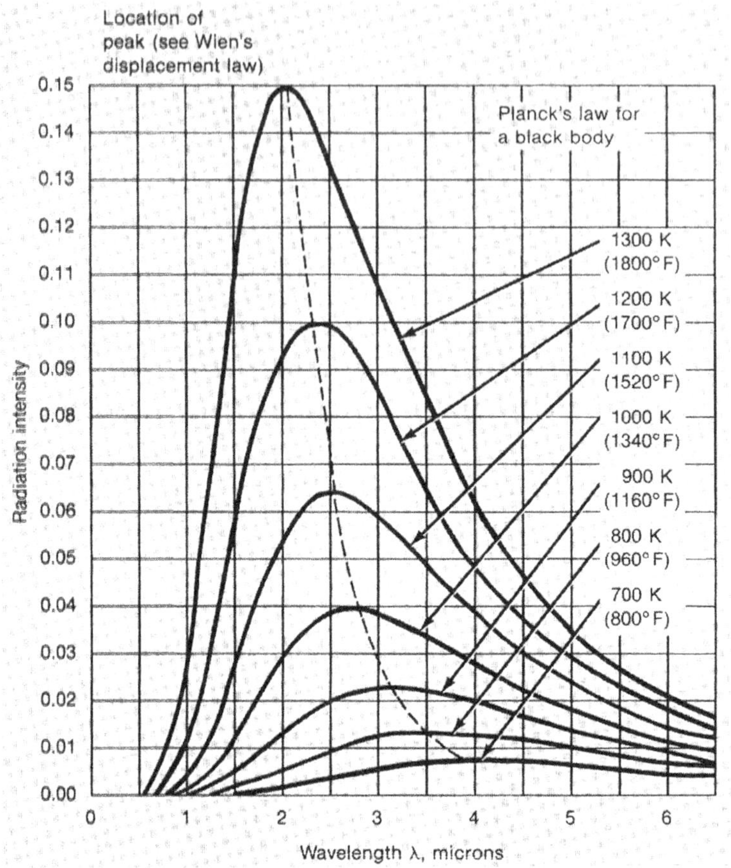

Figure 1-1. Blackbody radiation intensity as a function of wavelength and temperature *(Source: Figure 8-3. Temperature Measurement in Industry [ISA, 1990])*

tance depends upon application conditions and changes to the surface from coatings, corrosion, oxidation, and erosion during a production run.

Sensitivity

In Figure 1-1 for blackbodies the peak in the curve for each temperature shifts to the left as the temperature increases. The maximum change in radiation intensity for a change in temperature occurs near the peaks. To the left of the peak, the radiation intensity drops off suddenly. The change in intensity with temperature is appreciable until the intensity becomes close to zero. To the right of the peak, the radiation intensity tails off and the curves start to converge. The change in intensity with temperature becomes smaller and the pyrometer loses sensitivity. Figure 1-1 is for a blackbody [1][7, 8, 9, 10].

Graybodies

For graybodies, a fraction (e.g., 0.65 to 0.95) of the radiation intensity radiated by a blackbody is emitted. The fraction is fixed for all wavelengths for a given set of operating conditions. The shape of the graybody curve is similar to the blackbody curve and the peak would occur at the same wavelength but the graybody curve would be shifted below the blackbody curve as illustrated in Figure 1-2 by an amount that corresponds to the emittance of the graybody. The emittance can be a function of temperature and can vary with time but these changes are the same for all wavelengths. The graybody curve is simply the blackbody curve's values multiplied by the same emittance factor for all wavelengths. Equation 1-1c gives the total energy radiated for a graybody [1][7, 8, 9, 10].

Non-Graybodies

For non-graybodies, the fraction of the radiation emitted is variability and generally lower. The variation in emittance with wavelength and can result in multiple peaks as shown in Figure 1-2. The area under each of the peaks must be integrated to get the total radiation. We can get back to a simple general equation if we compute the spectral radiation (radiation at a specific wavelength) rather than the total radiation from all wavelengths. Equation 1-1d shows that a further modification of the Stephan-Boltzmann equation (that simply consists of raising the target temperature to the power N as a function of wavelength and temperature as defined by Equation 1-1e) provides a prediction of the spectral radiation for graybodies. If we make the emittance a function of wavelength besides temperature, we end up with Equation 1-1f, which provides a prediction of spectral radiation for non-graybodies.

Single and Two-color Pyrometers

In single-color and two-color pyrometers, wavelengths are chosen to provide a maximum change in radiation intensity with temperature but with minimum absorption by anything in the space between the target and the sensor. There is a peak in the absorption for the various gases, vapors, particles, and steam in the process equipment and the condition and material of the window into the process equipment. There is also a peak in the detector output. Ideally, the peak in radiation intensity and detector output coincide and are sufficiently separated from the peaks in absorption. As the maximum temperature increases, smaller wavelengths are desired because of the shift of the peak to the left in Figure 1-1 but this corresponds to peaks in absorption by gases, notably carbon dioxide.

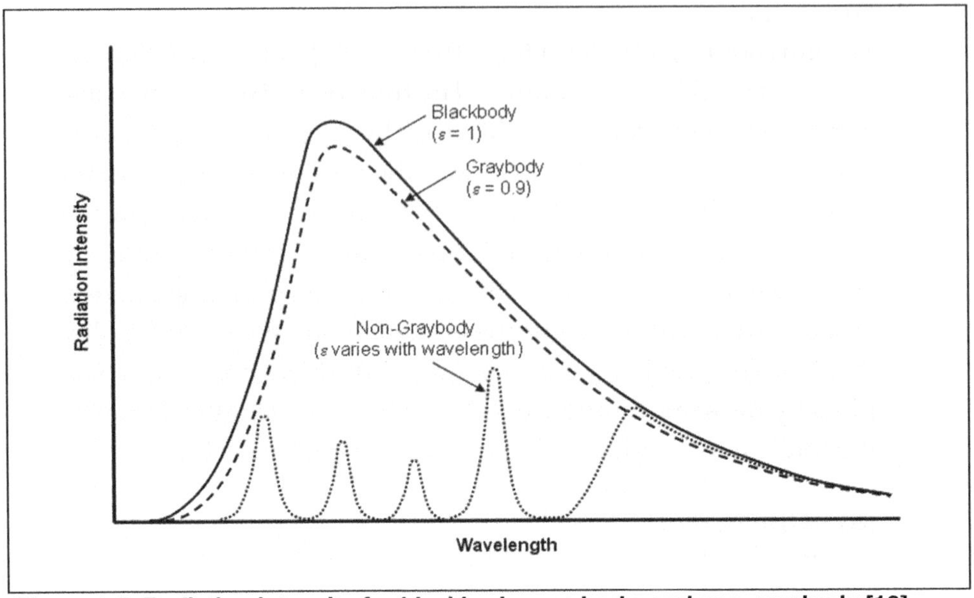

Figure 1-2. Radiation intensity for blackbody, graybody, and non-graybody [10]

Single-color or narrow band pyrometers use filters to narrow the view to a particular wavelength. In general a single-color pyrometer is more accurate than total radiation or broadband pyrometers (e.g., hand-held pyrometers) because smart instrumentation can take into account the known effects at a particular wavelength and a color can be chosen that is a better match for the application. However, there are often many unknown and time-varying effects, such as changes in the surface, the intervening vapors, and the viewing window, that introduce errors into the readings from a single-color pyrometer [1][7, 8, 9, 10].

Two-color or ratio pyrometers measure the radiation at two wavelengths. If the change in emittance at each wavelength with temperature is identical (graybodies), the effect of emittance can be cancelled out by ratio calculations. In reality, the change in emittance with temperature varies with wavelength (non-graybodies). Additionally, the change in emittance with changes in surface, operating conditions, and the composition of the intervening space may vary with wavelength.

In one comparison test on a blackbody, single-color and two-color pyrometers exhibited errors of 2 and 30°C (3.6 and 54°F), respectively [10]. Equal changes in emittance due to surface and operating conditions and intervening gases, particles, and vapors may make a two-color ratio pyrometer more accurate than a single-color pyrometer but it puts into question any accuracy statements for two-color pyrometers that are much better than

30°C. As a surface becomes closer to a blackbody and the absorption of radiation in the intervening space decreases, a single-color pyrometer may be best. The proper selection of the pyrometer type and wavelengths requires application experience or field trials.

The accuracy of a ratio two-color optical pyrometer may be less than a single color pyrometer because of the variation of emittance and radiation intensity with wavelength.

Equations for Total and Spectral Radiation

Total radiation intensity from a ***blackbody*** target can be expressed by the Stephan-Boltzmann Equation with the factor K to account for differences in target and sensor area and the distance between them:

$$E(T_t) = K * \sigma * (T_t^4 - T_s^4) \tag{1-1a}$$

For a negligible sensor temperature and a ***graybody*** with a target material emissivity that is a function of temperature, the equation for total radiation becomes:

$$E(T_t) = \varepsilon(T_t) * K * \sigma * (T_t^4) \tag{1-1b}$$

The substitution of target emittance for material emissivity takes into account installation and application effects on a ***graybody*** target on the total radiation:

$$E(T_t) = \kappa(T_t) * K * \sigma * (T_t^4) \tag{1-1c}$$

The substitution of N for the exponent of the ***graybody*** target temperature, gives the radiation intensity at a particular wavelength (spectral radiation):

$$E(\lambda_i, T_s) = \kappa(T_t) * K * \sigma * (T_t^N) \tag{1-1d}$$

$$N = \frac{14388}{\lambda_i * T_t} \tag{1-1e}$$

The spectral radiation intensities of ***non-graybodies*** have emittances that depend upon both wavelength and temperature:

$$E(\lambda_i, T_s) = \kappa(\lambda_i, T_t) * K * \sigma * (T_t^N) \qquad (1\text{-}1f)$$

where:

$E(T_t)$ = total radiation at a given temperature (Watts/cm^2)

$E(\lambda_l, T_t)$ = spectral radiation at a given wavelength and temperature (Watts/cm^2)

$\varepsilon(T_s)$ = emissivity of a surface at a given temperature

$\kappa(T_s)$ = emittance of a surface at a given temperature

$\kappa(\lambda_l, T_t)$ = emittance of a surface for a given wavelength and temperature

K = factor for distance and size of target and type of detector

σ = Stephan-Boltzmann constant (5.669x10^{-12} W/cm^{2}°K^{4})

T_t = target temperature (°K)

T_s = sensor temperature (°K)

λ_l = wavelength i (cm)

1-2. Thermocouples and RTDs

Sensing Element

The sensing element, which is constructed of metal, responds to the process temperature by generating a measurable resistance (RTDs) or voltage (TCs) signal.

Sensor Sheath

Most temperature sensors used in the process industry today have a sheath to prevent damage during handling and installation and from debris and solids. Furthermore, they are usually installed in a thermowell for additional protection and to meet the various material of construction requirements.

The sensor sheath, or cable housing, is constructed of metal and holds most of the component parts of the temperature sensor as shown in Figures 1-3a and 1-3b for an RTD. Typically magnesium oxide (MgO) sensor packing (also known as minerally insulated [MI]) surrounds the sensing element and is contained within the sensor sheath. The sensor packing decreases the impact of process vibration on the sensing element and thus ensures a more accurate measurement over time. The end of the sensor sheath is sealed with a fill (e.g., epoxy) that keeps moisture out of the sheath and away from the sensing element.

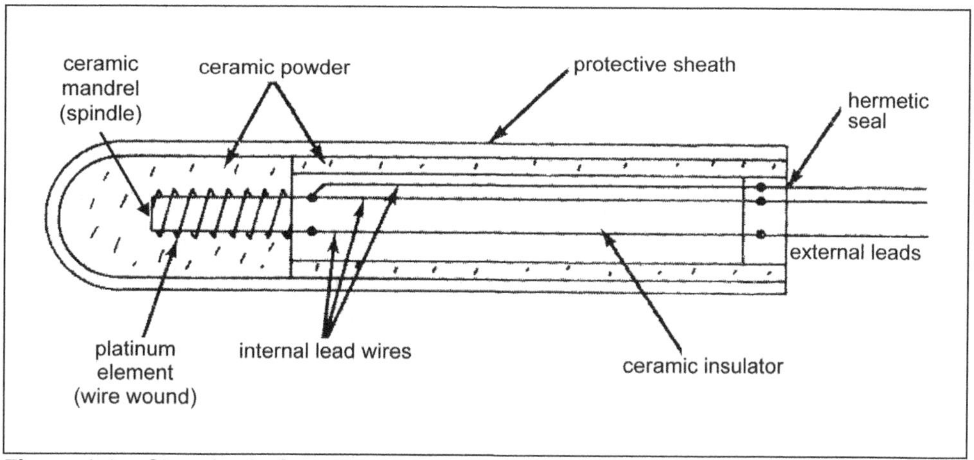

Figure 1-3a. Sheathed wire wound RTD sensor with 3 lead wires [8] (*Source: Figure 6-1. Temperature Measurement In Industry – [ISA, 1990]*)

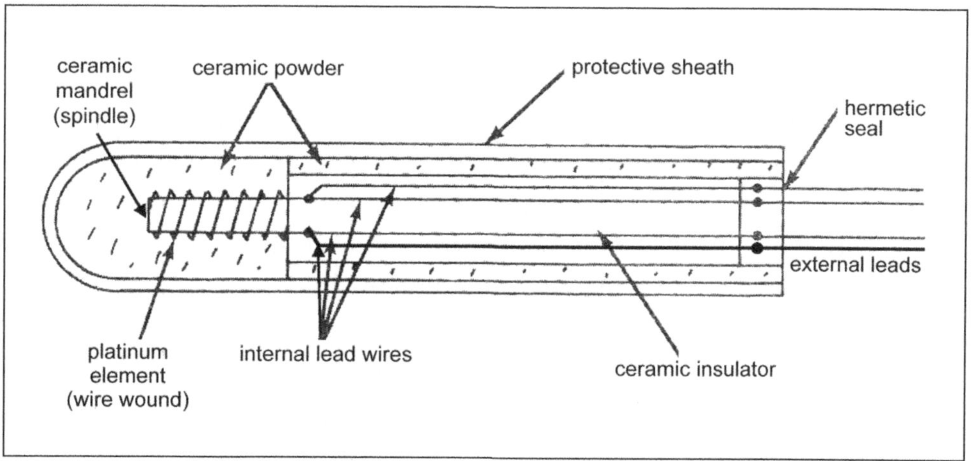

Figure 1-3b. Sheathed wire wound RTD sensor with 4 lead wires [8] (*Source: Figure 6-1. Temperature Measurement In Industry – [ISA, 1990]*)

Resistance Temperature Detectors (RTD)

RTDs operate on the principle that the electrical resistance of a metal increases as temperature increases, a phenomenon known as *thermoresistivity*. A temperature measurement can be inferred by measuring the resistance of the RTD element. The thermoresistive characteristics of RTD sensing elements vary depending on the metal or alloy from which they are made.

Platinum

Platinum RTD elements are the most common type used in process industries. Platinum elements have high accuracy, high repeatability, and a high resistance change per degree of temperature change. In addition, the output of platinum RTD elements is highly linear throughout their temperature range.

Copper

Copper RTD elements are highly linear throughout their temperature range, but have limited accuracy and a narrower temperature range than platinum elements. Copper elements, which are less expensive, are most often used for measuring temperature in bearings and motor windings—applications in which accuracy is not critical.

Wire-Wound RTD

Wire-wound RTD sensing elements are constructed by coiling a platinum (or other resistance metal) wire inside (internally wound) or around (externally wound) a ceramic *mandrel* (spindle). Most RTD sensors for the process industry are internally wound and sheathed for protection as shown in Figures 1-3a and 1-3b.

A dual-element wire-wound RTD can be created by coiling a second set of wires inside or outside the ceramic mandrel. If connected to a second transmitter, a transmitter with dual-sensor capabilities, or to another distributed control system (DCS) card, a *dual-element sensor* increases the reliability of the temperature measurement.

Wire-wound RTD elements are very sturdy and reliable. Compared to thin-film RTD elements, their accuracy tends to be higher, and their time response (how quickly the output reflects the temperature change) is several seconds faster than thin-film RTD elements. Wire-wound RTD elements work well for a wide variety of applications, although they may fail in high-vibration applications. The single element has a lower gauge sensing element and smaller time constant than the dual element. The use of redundant sensors helps eliminate common mode failures and enables a better cross check of sensor drift than dual elements. Three sensors and middle signal selection reduces noise and drift and provides inherent automatic protection against a single failure of any type.

Two single-element RTD sensors, instead of a dual-element sensor, provide a faster response with better diagnostics (such as drift) and less susceptibility to vibration and common mode failures.

Thin-Film RTD

Thin-film RTD sensing elements are constructed by depositing a thin film of resistance metal onto a ceramic substrate (base piece) and trimming the metal to specifications. Sensing elements of thin-film construction are typically less expensive than those of wire-wound construction because less resistance metal is required for construction. However, thin-film RTDs tend to be less stable over time, typically have a more limited temperature range, and may be more susceptible to damage from rough handling.

Extension Lead Wires

To get an accurate temperature reading from an RTD, the resistance of the RTD sensing element must be measured. Each copper lead wire that connects the RTD sensing element to the resistance measuring device adds a small amount of resistance to the measurement. If this added resistance is ignored, an error is introduced and an inaccurate temperature measurement results. The error is referred to as the *lead wire effect*. The longer the wire run, the greater the error, or lead wire effect, reflected in the temperature measurement. To compensate for lead wire effect, three-wire and four-wire RTDs are used instead of two-wire RTDs. Three-wire RTDs are created by connecting one additional copper wire to one of the lead wires. Four-wire RTDs are created by connecting one additional copper lead wire to each of the existing lead wires. These additional wires are used by the transmitter to compensate for lead wire resistances.

The third wire compensates for the resistance of the lead wires based on the assumption that each wire has exactly the same resistance. In fact, there is a tolerance of 10% in the resistance of standard wires. The fourth wire compensates for the uncertainty in the resistance of wires. For example, 500 feet of 20 gauge cable would add 10 ohms which would cause a measurement error of 26°C (47°F) for a two-wire RTD. The 10% tolerance of the cable could create an error as large as 2.6°C (4.7°F) for a three-wire RTD [11]. For high accuracy applications or long extension wire runs, a four-wire RTD or a transmitter mounted on the thermowell (integral mount) should be used. The increased accuracy, stability, and reliability of microprocessor based transmitters and the advent of secure and reliable wireless networks makes integral mounted transmitters an attractive option. Accessibility is less of an issue because maintenance requirements are drastically reduced. The transmitters rarely need removal, wiring problems are gone, and calibration checks and integrity interrogation can be done remotely.

A four-lead wire RTD is advisable for tight measurement accuracy requirements or long distances because the fourth wire can compensate for the differences in lead wire resistance not compensated by a three-lead wire RTD.

Once the resistance value is determined, it is converted to a temperature measurement. One of two conversion methods may be used by the transmitter:

- RTD standard (i.e., IEC 751 standard)
- Callendar-Van Dusen equation

IEC 751 standard

The IEC 751 standard describes an ideal relationship between the resistance of a platinum RTD and the temperature to which the RTD is subjected. For example, at 100°C (212°F), the IEC 751 standard shows that an ideal platinum RTD (one that exactly matches the IEC 751 standard) would have a resistance value of 138.5 ohms.

When a transmitter or control system accepts a resistance signal from a platinum RTD, the IEC 751 standard curve is often used to translate that resistance signal into a temperature reading. However, since actual RTDs are never ideal, they do not match the resistance versus temperature relationship as described in the IEC 751 standard. The difference between the actual RTD curve and the ideal RTD curve results in a measurement error, which is referred to as a *sensor interchangeability error.*

The maximum allowable sensor interchangeability error at a given process temperature is defined by two IEC 751 standard classifications—Class A and Class B. These classifications are used to identify platinum RTDs. Figure 1-4 compares these two classes to the IEC 751 standard ideal.

Callendar-Van Dusen Equation

The Callendar-Van Dusen equation offers an alternative to the IEC 751 standard. The equation, used in "Transmitter-Sensor Matching," can cre-

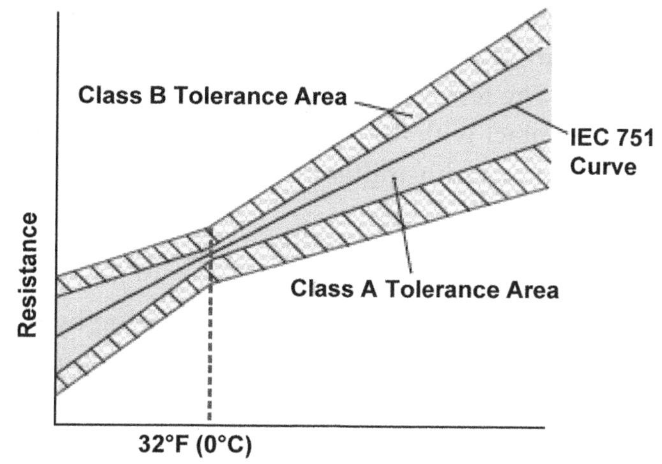

Figure 1-4. IEC 751 standard

ate a curve that approximates an RTD's resistance-temperature relationship. The Callendar-Van Dusen equation is:

$$R_t = R_O + R_O\alpha[(t-\delta)(0.01t-1)(0.01t) - \beta(0.01t-1)(0.01t)^3]$$

where:

t = Temperature in °C

R_t = Resistance of the RTD at t

R_O= Resistance of the RTD at t = 0°C (a Callendar-Van Dusen constant)

α, β and δ = Callendar-Van Dusen constants

The Callendar-Van Dusen equation can be programmed into a transmitter so that the transmitter can use the actual RTD curve rather than an ideal curve (e.g., IEC 751 standard) to translate the sensor's resistance signal into a temperature value. The Callendar-Van Dusen equation provides a significant improvement in measurement accuracy, even when compared to Class A RTDs.

The Callendar Van Dusen equation can eliminate most of the RTD sensor interchangeability error.

Thermocouples (TC)

A thermocouple (TC) consists of two wires of dissimilar metals (e.g., iron and constantan) that are joined at one end to form a *hot junction* (or sensing element). The temperature measurement is made at the hot junction, which is in contact with the process. The other end of the TC lead wires, when attached to a transmitter or volt meter, forms a *cold* or *reference* junction.

Thermocouple Types

Several types of TCs are available, each differing by the metals used to construct the element. While accuracies are better for type T and E compared to J, the type selected in industry often comes down to the plant standards and the application temperature range. Figure 1-5 shows the milliVolts generated over the approximately linear range of the following types of thermocouples:

- Type E—Chromel and constantan
- Type J—Iron and constantan
- Type K—Chromel and alumel
- Types R and S—Platinum and rhodium (differing in the % of platinum)
- Type T—Copper and constantan

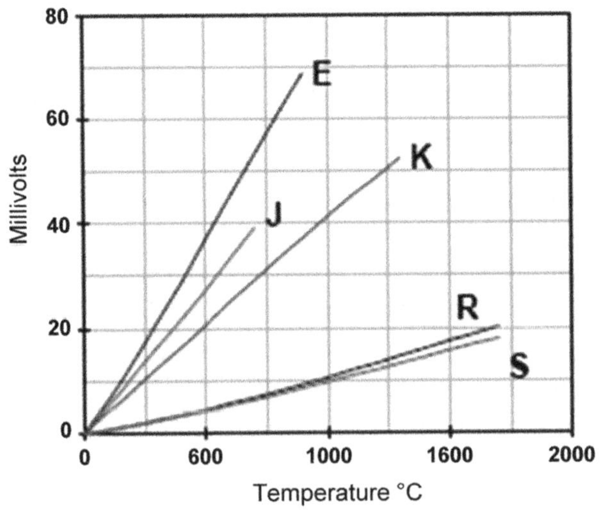

Figure 1-5. MilliVolts generated by various thermocouple types

Hot Junction Configurations

Junctions can be grounded or ungrounded to the sensor sheath. With dual-element TCs (two TCs in one sheath), the elements can be isolated or connected (unisolated). Each configuration offers benefits and limitations:

- *Grounded*—Grounding creates improved thermal conductivity, which in turn gives the quickest response time. However, grounding also makes TC circuits more susceptible to electrical noise (which can corrupt the TC voltage signal) and may cause more susceptibility to poisoning (contamination) over time.
- *Ungrounded*—Ungrounded junctions have a slightly slower response time than grounded junctions, but are not susceptible to electrical noise.
- *Unisolated*—Unisolated junctions are at the same temperature, but both junctions will typically fail at the same time.
- *Isolated*—Isolated junctions may or may not be at the same temperature. The reliability of each junction is increased, because failure of one junction does not necessarily cause a failure in the second junction.

Grounded thermocouples provide the fastest and most accurate response by minimizing dynamic error and the temperature difference from the process by minimizing the thermal resistance between the sensor and the process. Ungrounded thermocouples are less susceptible noise and contamination (see Chapter 2 for more details).

Voltage Measurement and the Seebeck Effect

TCs use a phenomenon known as the Seebeck effect to determine process temperature. According to the Seebeck effect, a voltage measured at the cold junction of a TC is proportional to the difference in temperature between the hot junction and the cold junction. The voltage measured at the cold junction is commonly referred to as the Seebeck voltage, the thermoelectric voltage, or the thermoelectric EMF. As the temperature of the hot junction (or process fluid) increases, the observed voltage at the cold junction also increases by an amount nearly linear to the temperature increase.

If the hot junction temperature is held constant, an increasing cold junction temperature will produce a decreasing voltage, because the temperature difference between the hot and cold junction is decreasing. When the cold and hot junctions reach the same temperature, the observed voltage will be 0 V. The magnitude of the voltage signal produced at the cold junction also depends on the type of metals used to form the TC. Different combinations of metals, or different TC types, have different thermoelectric voltages at the same temperatures.

Cold Junction Compensation

As with RTDs, each type of TC has a standard curve. The standard curve describes a TC's voltage versus temperature relationship when the cold junction temperature is 0°C (32°F). As mentioned, the cold junction is where the TC lead wires attach to a transmitter or volt meter. Because the voltage measured at the cold junction is proportional to the difference in temperature between the hot and cold junctions, the cold junction temperature must be known before the voltage signal can be translated into a temperature reading. The process of factoring in the actual cold junction temperature (rather than assuming it is at 0°C [32°F]) is referred to as cold junction compensation (CJC).

1-3. Selection of a Temperature Sensor

Why Use an RTD Rather Than a TC?

The main reasons for selecting RTDs rather than TCs are as follows:

- RTDs have better sensitivity, repeatability, and stability.
- RTD signals are less susceptible to noise (higher signal-to-noise ratio).
- RTDs have better linearity over temperature ranges.
- RTDs can use the Callendar-Van Dusen equation to eliminate sensor interchangeability error.
- Cold junction compensation and related errors are not associated with RTDs.
- RTD drift is predictable, while TC drift is erratic and unpredictable. In addition, TC drift errors can be large as a result of element poisoning and element oxidation at high temperatures.
- The changes that affect the output of an RTD or TC occur over time due to mechanical shock, poisoning, and temperature

cycling. These changes can be eliminated by an in-line RTD calibration, an option not available for a TC.

- RTDs do not need special extension wire.

Why Use a TC Rather Than an RTD?
In summary, the main reasons for selecting TCs rather than RTDs are as follows:

- TCs function at higher temperatures than RTDs (above 850°C [1500°F]).
- TCs are typically less expensive than RTDs and more resistant to vibration damage but the life cycle costs are higher.
- TCs have a faster response time than RTDs but this is only a consideration for bare elements (direct immersion of the element without a thermowell) in very fast temperature change processes (process time lag < 10 seconds). Bare elements are not used in industrial chemicals or elevated temperatures due to safety concerns.

The main reasons to use a TC instead of a RTD are high temperatures and high vibration.

1-4. Specifications

The accuracy of a temperature measurement is based on the combination of the sensor used and the performance specifications of the transmitter. The accuracy ranges shown in Table 1-3 are what can be expected for various sensor/transmitter combinations.

If supported by the transmitter, sensor matching using the Callendar-van Dusen constants for an RTD sensor will improve the accuracy by 75%. Ambient temperature effect: 0.001% of span or < 0.007°C per 1°C (0.0013°F per 1°F) of temperature change for most sensor types. Stability will range from ±0.25% of reading for 5 years for RTDs to ± 0.5% of reading for 1 year for thermocouples.

Table 1-3. Range and accuracy specifications

RTD Sensor Type	RTD Sensor Reference	Sensor Range		Accuracy over Range	
		°C	°F	°C	°F
Pt 100 (_ = 0.00385)	IEC 751; _ = 0.00385, 1995	–200 to 850	–328 to 1562	± 0.10 to ± 0.30	± 0.18 to ± 0.54
Pt 100 (_ = 0.003916)	JIS 1604, 1981	–200 to 645	–328 to 1193	± 0.10 to ± 0.30	± 0.18 to ± 0.54
Pt 200	IEC 751; _ = 0.00385, 1995	–200 to 850	–328 to 1562	± 0.22 to ± 0.54	± 0.40 to ± 0.98
Pt 500	IEC 751; _ = 0.00385, 1995	–200 to 850	–328 to 1562	± 0.14 to ± 0.38	± 0.25 to ± 0.68
Pt 1000	IEC 751; _ = 0.00385, 1995	–200 to 300	–328 to 572	± 0.10 to ± 0.40	± 0.18 to ± 0.72
Ni 120	Edison Curve No. 7	–70 to 300	–94 to 572	± 0.10 to ± 0.30	± 0.18 to ± 0.54
Cu 10	Edison Copper Winding No. 15	–50 to 250	–58 to 482	± 1.00 to ± 3.20	± 1.80 to ± 5.76
Cu 100 (a=428)	GOST 6651-94	–185 to 200	–365 to 392	± 0.48	± 0.86
Cu 50 (a=428)	GOST 6651-94	–185 to 200	–365 to 392	± 0.96	± 1.73
Cu 100 (a=426)	GOST 6651-94	–50 to 200	–122 to 392	± 0.48	± 0.86
Cu 50 (a=426)	GOST 6651-94	–50 to 200	–122 to 392	± 0.96	± 1.73

TC Sensor Type	TC Sensor Reference	Sensor Range		Accuracy over Range	
		°C	°F	°C	°F
NIST Type B	NIST Monograph 175	100 to 300	212 to 572	± 3.00 to ± 6.00	± 5.40 ±10.80
(varies by input range)		301 to 1820	573 to 3308	± 0.75 to ± 1.54	± 1.35 to ± 2.78
NIST Type E	NIST Monograph 175	–50 to 1000	–58 to 1832	± 0.20 to ± 0.40	± 0.36 to ± 0.72
NIST Type J	NIST Monograph 175	–180 to 760	–292 to 1400	± 0.25 to ± 0.70	± 0.45 to ± 1.26
NIST Type K	NIST Monograph 175	–180 to 1372	–292 to 2502	± 0.25 to ± 1.00	± 0.45 to ± 1.80
NIST Type N	NIST Monograph 175	–200 to 1300	–328 to 2372	± 0.40 to ± 1.00	± 0.72 to ± 1.80
NIST Type R	NIST Monograph 175	0 to 1768	32 to 3214	± 0.60 to ± 1.50	± 1.08 to ± 2.70
NIST Type S	NIST Monograph 175	0 to 1768	32 to 3214	± 0.50 to ± 1.40	± 0.90 to ± 2.52
NIST Type T	NIST Monograph 175	–200 to 400	–328 to 752	± 0.25 to ± 0.70	± 0.45 to ± 1.26
DIN L	DIN 43710	–200 to 900	–328 to 1652	± 0.35 to ± 0.70	± 0.63 to ± 1.26
DIN U	DIN 43710	–200 to 600	–328 to 1112	± 0.35 to ± 0.70	± 0.63 to ± 1.26
w5Re26	ASTME 988-96	0 to 2000	32 to 3632	± 0.70 to ± 1.60	± 1.26 to ± 2.88
GOST Type L	GOST R 8.585-2001	–200 to 800	–392 to 1472	± 0.71	±1.28

1-5. Setup and Calibration

Calibration is a means to verify the proper operation of the sensor and the electronics receiving the sensor signal.

Sensors can be calibrated by exposing them to one or more known temperatures and measuring the sensor output. The output is then compared to that of a certified sensor probe to determine the calibration error. It is also possible to calibrate the combined measurement system by measuring the resulting reading of the transmitter or other measurement electronics.

Transmitters and other electronics measurement equipment such as system I/O cards can be calibrated by applying an electronically generated signal to simulate the voltage or resistance of a sensor. A wide variety of calibrators are available for this purpose.

Most suppliers of temperature sensors and transmitters will provide factory calibration for their products with associated documentation of the results.

1-6. Installation

The best practice for making a temperature measurement is to keep the length of the sensor wiring as short as possible to minimize the effect of electromagnetic interference (EMI) and other interference on the low level sensor signal. The temperature transmitter should be mounted as close to the process connection as possible.

To minimize conduction error (error from heat loss along the sensor sheath or thermowell wall from tip to flange or coupling), the immersion length should be at least 10 times the diameter of the thermowell or sensor sheath for a bare element. Thus, for a thermowell with a 1 inch (2.54 cm) outside diameter, the immersion length should be 10 inches (25.4 cm). For a bare element with a ¼ inch (6.35 mm) outside diameter sensor sheath, the immersion length should be at least 2.5 inches (63.5 mm). This is just a rule of thumb. Computer programs can compute the error and do a fatigue analysis for various immersion lengths and process conditions.

For high velocity stream and bare element installations, it is important to do a fatigue analysis because the potential for failure from vibration increases with immersion length.

The process temperature will vary with process fluid location in a vessel or pipe due to imperfect mixing and wall effects. For highly viscous fluids such as polymers and melts flowing in pipes and extruders, the fluid temperature near the wall can be significantly different than at the centerline (e.g., 10 to 30°C; 50 to 86°F) [12]. Often the pipelines for specialty polymers are less than 4 inches (101.6 mm) in diameter, presenting a problem for getting sufficient immersion length and a centerline temperature measurement. The best way to get a representative centerline measurement is by inserting the thermowell in an elbow facing into the flow (position 1 in Figure 1-6). If the thermowell is facing away from the flow, swirling and separation from the elbow as can create a noisier and less representative measurement (position 2 in Figure 1-6). An angled insertion (position 3 in Figure 1-6) can increase the immersion length over a perpendicular insertion (position 4 in Figure 1-6) but the insertion lengths shown for both are too short unless the tip extends past the centerline. A swaged or stepped thermowell can reduce the immersion length requirement by reducing the diameter near the tip.

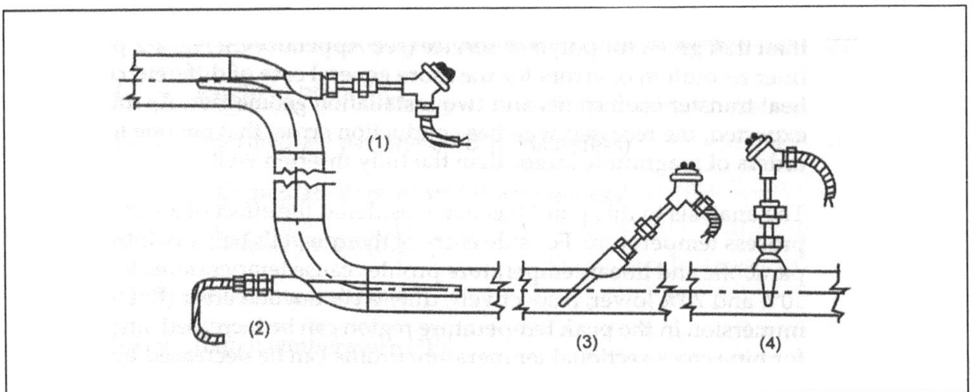

Figure 1-6. Ranking of installations to reduce heat conduction and profile errors (*Source: 614. Advanced Control Unleashed [ISA, 2003]*)

Swaged and tapered thermowells in elbows reduce the thermowell length requirement and the pipe temperature profile error.

The distance of the thermowell in a pipeline from a heat exchanger, static mixer, or desuperheater outlet should be optimized to reduce the transportation delay but minimize noise from poor mixing or two phase flow. As shown in Figure 1-7, generally 25 pipe diameters is sufficient to ensure sufficient mixing after the recombination of divided flows from heat exchanger tubes or static mixer elements.

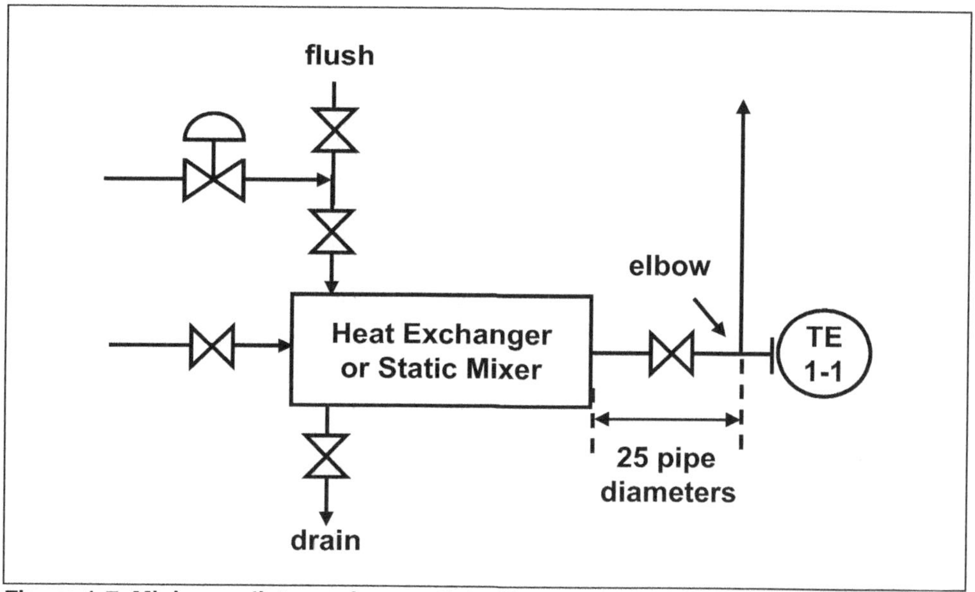

Figure 1-7. Minimum distance from outlet of heat exchanger and static mixer

For desuperheaters, the distance from the outlet to the thermowell depends upon the performance of the desuperheater, process conditions, and the steam velocity. To give a feel for the situation there are some simple rules of thumb for the straight piping length (SPL) to the first elbow and the total sensor length (TSL). Actual SPL and TSL values depend on the quantity of water required with respect to the steam flow rate, the temperature differential between water and steam, the water temperature, pipe diameter, steam velocity, model, type, etc. and are computed by software programs [13].

SPL (feet) = Inlet steam velocity (ft/s) * 0.1 (seconds residence time)
SPL (m) = Inlet steam velocity (m/s) * 0.1 (seconds residence time)

TSL (feet) = Inlet steam velocity (ft/s) * 0.2 (seconds residence time)
TSL (m) = Inlet steam velocity (m/s) * 0.2 (seconds residence time)

Typical values for the inlet steam velocity, upstream of the desuperheater range from 25–350 ft/s (7.6 to 107 m/sec). Below 25 ft/s there is not enough motive force to keep the water suspended in the steam flow. Water tends to fall out and run down the pipe to a drain. When this happens the water no longer cools the steam and the system thinks it needs to add more water, which compounds the problem. Problems can also include pipe wall erosion and high thermal stress gradients in the pipe wall (i.e., a hot top and cold bottom, which can crack welds or warp the pipe to an egg-shaped cross-section). Current technology has an inlet velocity limitation of 350 ft/s (107 m/sec). Velocities higher than 350 ft/s cause the desuperheater to vibrate and damage the unit to the point where it breaks apart [13].

For desuperheaters, the straight pipe length to the first elbow and the total length to the sensor from the desuperheater outlet must provide a residence time of 0.1 and 0.2 seconds, respectively, to prevent water droplets from hitting sensor.

Thermowells

As mentioned, a *thermowell* is a closed-end metal tube that encapsulates and protects a sensor from process flow, pressure, vibration, and corrosion. Thermowells allow for the installation and removal of sensors without having to shut down the process. They are mounted in various ways to a process pipe or vessel.

Thermowells are available in several different materials, mounting methods, and stem types. The variety of design features renders thermowells suitable for various applications and environmental conditions. Three factors affect the choice of material:

- Type of corrosive environment the thermowell will be exposed to
- Temperature and pressure limits of the material
- Compatibility with the process piping or vessel material to ensure solid, non-corroding welds and junctions

Mounting Methods

Thermowells and sensing elements can be assembled and mounted as shown in Figures 1-8 through 1-9. Thermowells can be threaded, welded, or bolted (flanged style) onto the process pipe or vessel wall process connection.

Thermowells are threaded onto the process piping or vessel, which enables them to be easily installed and removed. Threaded thermowells are the weakest type of thermowell.

Welded thermowells are permanently welded onto the process pipe or vessel. Thus, removal is very difficult and requires cutting the thermowell out of the system. Welded thermowells are the strongest type of thermowell and are used with fluids of high velocity, high temperature, or high pressure. Welded thermowells are necessary for applications that require a leak-proof seal.

Flanged thermowells are bolted onto a pipe or vessel and can be easily removed or installed. Flanged thermowells are used in corrosive environments, as well as in high-velocity, high-temperature, or high-pressure applications. Flanged thermowells are the most expensive type of thermowell.

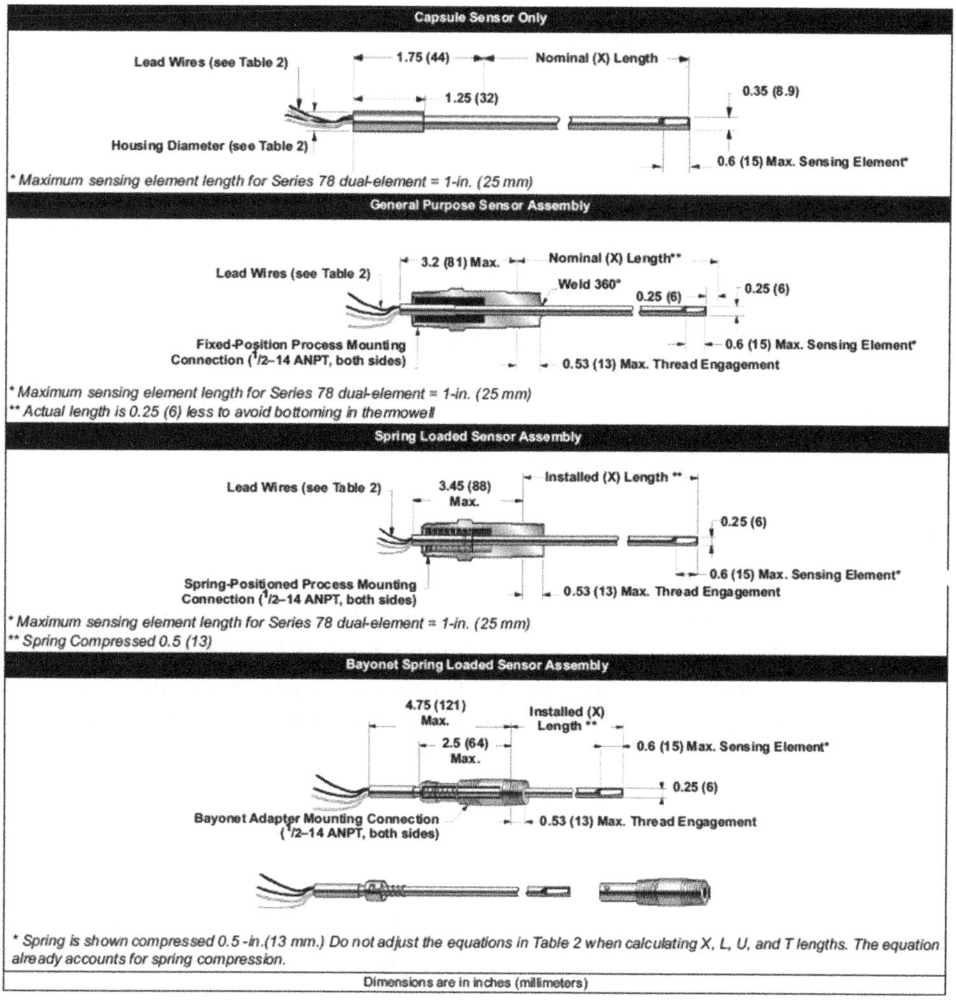

Figure 1-8. RTD sensor assembly types

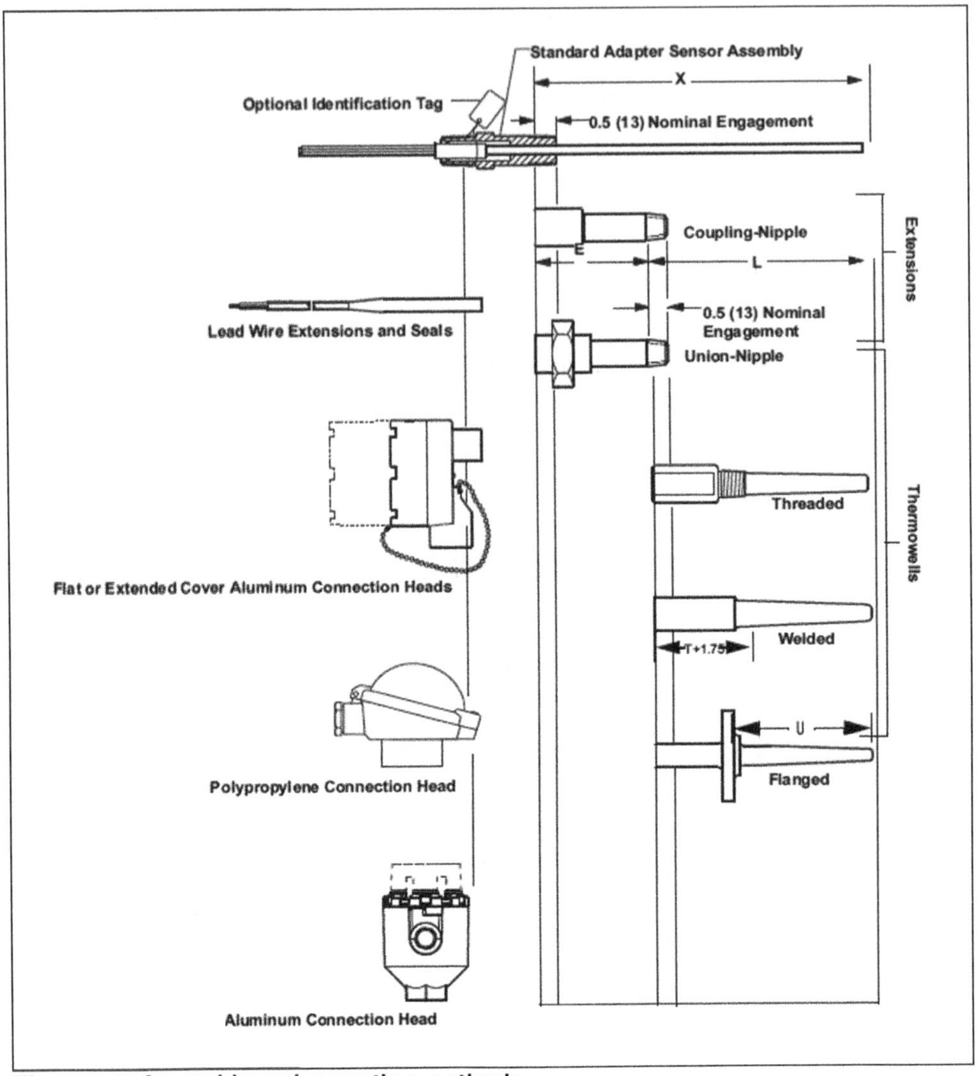

Figure 1-9. Assembly and mounting methods

> *Flanged thermowells are commonly used in chemical plants because of concern of the long-term integrity of threaded connections for corrosive and harsh process conditions. Hazardous chemicals and high pressures and temperatures may require welded connections.*

The mounting of the transmitter on the thermowell as shown in Figure 1-10 eliminates the need for cables between the element and transmitter and offers mobility for wireless transmitters important for finding the optimum temperature measurement location.

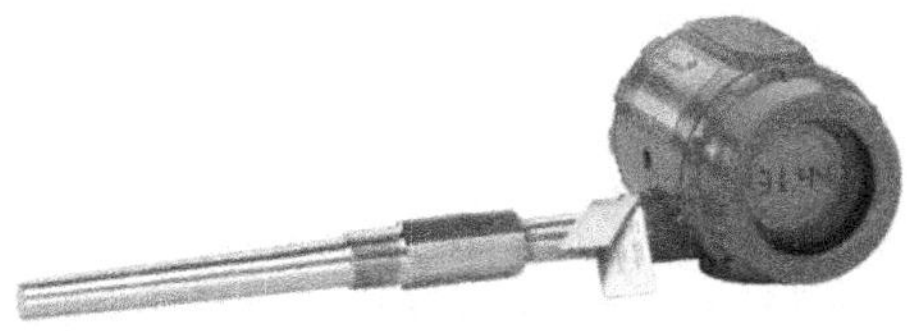

Figure 1-10. Transmitter mounted on thermowell

Stem Designs

The *stem* of a thermowell is the part that is inserted into the process stream. Stems can be tapered, straight, or stepped. The performance of a thermowell varies with its stem design. In general, a tapered or stepped stem provides a faster response, creates less pressure drop, and is less susceptible to conduction error and vibration failure. The choice of stem design is based on:

- Process pressure
- Time response required
- Permissible conduction error
- Wake frequency—frequency of alternating side-to-side movement of a fluid; depends on properties of the fluid
- Drag force—resistance to motion of a solid shape through a body of fluid
- Price

If the thicknesses of the thermowell walls and the fit of the sensing element are identical, thermowells with straight stems have the slowest time response because they possess the most material at the tip (largest diameter). Thermowells with stepped stems have the fastest time response because they possess the least material at the tip (smallest diameter). A small diameter also results in the least amount of drag force.

Thermowells with stepped stems also provide the maximum separation between the wake frequency (vortex shedding) and the natural frequency (oscillation rate determined by the properties of the thermowell itself). If the wake frequency is 80% or more of the thermowell natural frequency, resonance and probably damage can occur. Generally, thermowells with tapered stems are slightly more expensive as a result of a more complicated manufacturing process.

Swaged, stepped, and tapered thermowells offer a faster response, lower pressure drop, and less possibility of vibration damage from resonance with wake frequencies.

Communication

To be useful for control, safety, or monitoring applications, a temperature measurement signal must be communicated from the point of measurement to the control system of the process. The two most common ways are:

- *Transmitter*—the sensor is wired a short distance to the transmitter or connected directly to the transmitter, where its signal is converted to a digital, 4–20 mA, or wireless signal. The converted signal output is then communicated to the control system through transmitter wire or a wireless network.
- *Wired direct*—the sensor's lead wires are wired the entire distance to the control system. No signal conversion takes place along the route.

Three benefits of using a temperature transmitter over wiring directly to thermocouple and RTD input cards of control system are:

- *A more robust signal is delivered*—the 4–20 mA or digital signal output from the transmitter is much more robust than a sensor signal that is wired direct. Noise interference has less impact on 4–20 mA or digital signals.
- *Measurement accuracy is optimal*—Transmitters offer improved measurement accuracy over wiring direct. For example, sensors can be matched to transmitters (transmitter sensor matching), which improves the accuracy of the temperature measurement. The temperature span can be narrowed to match the process operating range (significant for older DCS with 12-bit input cards).
- *Time and money are saved*—Transmitter installation is often less expensive than wiring a sensor direct because of savings from cabling costs and installation (sensor wire, especially TC wire, is relatively expensive). Also, a robust signal and accurate mea-

surements produce time and money savings through increased functionality and diagnostic capabilities of the transmitter.

Typically, the analog signal is linear with a process temperature measurement. Most transmitters now incorporate a microprocessor, which has improved their performance compared to analog designs. These smart (also called "intelligent") transmitters are able to compensate for ambient temperature variations and EMI, and provide cold junction compensation for thermocouples. Some also enable the transmitter to be matched to the characteristics of a specific sensor, providing very high accuracy. They are easily configured for a specific sensor type, providing an output that is linear over the temperature range. Intelligent transmitters with digital communications are also able to communicate diagnostic information about the health of the sensor and electronics.

Some transmitters provide the ability to connect more than one sensor. The most common transmitter of this type will support two sensors but has a single analog output. Coupled with dual sensor elements, this feature can be used to provide a more reliable measurement by comparing two measurements or the ability to switch to the second sensor in the event of the failure of the first.

The use of transmitters instead of TC or RTD input cards is recommended to greatly improve accuracy and maintainability, by matching the calibration and nonlinearity compensation to sensor, narrowing the span, reducing noise, and offering diagnostics.

The use of digital communications allows the additional flexibility of using a single transmitter to make more than one temperature measurement and communicate these back to the control system. There are temperature devices designed to specifically take advantage of this capability, providing the ability to measure four, or eight, or potentially more individual temperatures. The most common communication techniques are the HART® (including WirelessHART™), FOUNDATION™ Fieldbus and Profibus PA standard protocols.

The reliability, security, and ease of setup of WirelessHART (Highway Addressable Remote Transducer) networks combined with increased battery life from new communication rules and PID enhancements have made wireless communication an excellent option [14]. Since temperature changes in most processes are quite slow, the refresh time can be set longer than for other types of loops, extending battery life. Also, the noise amplitude and period in temperature loops is usually quite small compared to other loops unless there are two phases (e.g., liquid and gas) or poor mixing (e.g., poor uniformity—increased variability due to insufficient agitation), decreasing the number of exception updates triggered by noise, which also extends battery life.

Field-Mount Transmitters

Field-mount transmitters are the most rugged of all transmitter styles. Their robust housings protect against corrosion and humidity. Some field-mount transmitters house the electronics in dual-compartment housings, which completely isolates them from the effects of humidity. Dual-compartment transmitters are the best design for use in harsh environments. Field-mount transmitters can be integrally or remotely mounted.

The use of integral mounting and wireless transmitters provides portability for monitoring unit operation efficiency and finding the most representative and sensitive measurement location with the least process dead time.

- *Integral mount*—the transmitter is threaded onto the sensor directly (mounts directly to U.S. style [1/2-inch NPT] sensor or DIN plate).
- *Remote mount*—the transmitter is mounted on a pipe stand or other support near the sensor. Remote mount is preferred when the measurement point is inaccessible or when the process environment is too harsh for the transmitter to be installed directly on top of the sensor.

The integral mounting of a wireless transmitter, where permitted by accessibility and temperature, enables portability for online process and equipment performance metrics and optimization of measurement location.

Head-Mount Transmitters

Head-mount transmitters are small, puck-shaped transmitters. They are typically housed in a protective enclosure—a *connection head* for direct mounting or a *junction box* for remote mounting.

Rail-Mount Transmitters

Rail-mount transmitters are designed to be attached to a DIN-rail (G-rail or top-hat rail) or directly screwed onto a wall. Rail-mount transmitters are also designed for compact mounting, which allows for a number of transmitters to be mounted very closely together

Wiring Direct

As mentioned, *wiring direct* refers to wiring the sensor's lead wires back to the control system. Because the sensor's lead wire (and original signal) is traveling the entire distance from the point of measurement to the control system, care must be taken to avoid two key problems:

- *Noise*—TCs are especially sensitive to noise interference and extension wires must be routed around such sources as generators and motors.
- *Heat sources*—a large change in the ambient temperature can affect the sensor's signal as it travels to the control system.

Extension Wire

TC extension wires are often used to wire a TC back to a control system or to a remote transmitter, which may be anywhere from 2–2000 ft (0.61–610 m) away. TC extension wire, with a few rare exceptions, must be of the same type of metal as the TC lead wires. If the metals do not match, the cold junction will be created prematurely. TC wire is relatively costly, particularly for platinum TC Types R and S. It is often not economically feasible to make continuous runs of TC wire (perhaps hundreds of feet long) from the hot to the cold junctions. To compensate for this problem, special lead wire is used that closely approximates the thermoelectric properties of TC wire. The special wire with less expensive material allows the user to minimize cost without sacrificing performance.

1-7. Maintenance

Table 1-4 lists checks that can be made to manually track down the source of a high, erratic, or low transmitter output.

The use of intelligent transmitters can simplify the process of diagnosing problems, since they are continuously monitoring both the sensor and the electronics. Transmitter diagnostics typically include indication of an open or shorted sensor. They can also detect if the measurement is outside of the sensor range or beyond the range limits of the analog output.

As mentioned, more advanced transmitters add diagnostic capabilities such as the ability to have a second sensor input back up the first in the event of failure, and also to compare the inputs from two sensors to determine if one is drifting. Advanced diagnostics are available that can detect if a sensor's performance is deteriorating even before the condition affects the measurement, allowing the sensor to be replaced or the wiring corrected before the process is impacted.

Table 1-4. Temperature measurement troubleshooting guide example

High Output	Sensor Input	• Check for a sensor open circuit. • Check if the process variable is out of range.
	Loop Wiring	• Check for dirty or defective terminals, interconnecting pins, or receptacles.
	Power Supply	• Check the output voltage of the power supply at the transmitter terminals. It should be 12.0 to 42.4 V DC (over entire 3.5- to 23.0-mA operating range).
	Electronics Module	• Connect a Field Communicator and enter the transmitter test mode to isolate module failure. • Connect a Field Communicator and check the sensor limits to ensure calibration adjustments are within the sensor range.
Erratic Output	Loop wiring	• Check for adequate voltage to the transmitter. It should be 12.0 to 42.4 V DC at the transmitter terminals (over entire 3.5- to 23.0-mA operating range). • Check for intermittent shorts, open circuits, and multiple grounds. • Connect a 375 Field Communicator and enter the loop test mode to generate signals of 4 mA, 20 mA, and user-selected values.
	Electronics Module	• Connect a Field Communicator and enter the transmitter test mode to isolate module failure.
Low Output or No Output	Sensor Element	• Check if the process variable is out of range.
	Loop Wiring	• Check for adequate voltage to the transmitter. It should be 12.0 to 42.4 V DC (over entire 3.5- to 23.0-mA operating range). • Check for shorts and multiple grounds. • Check for proper polarity at the signal terminal. • Check the loop impedance. • Connect a Field Communicator and enter the loop test mode. • Check wire insulation to detect possible shorts to ground.
	Electronics Module	• Connect a Field Communicator and check the sensor limits to ensure calibration adjustments are within the sensor range. • Connect a Field Communicator and enter the transmitter test mode to isolate an electronics module failure.

Exercises

1-1. What are the accuracy advantages of an RTD?

1-2. What are the advantages of thermocouples?

1-3. When is a Thermistor used?

1-4. Where are non-contacting optical pyrometers used?

1-5. What affects the indication of an optical pyrometer besides target temperature?

1-6. When is a 2-color pyrometer effective?

1-7. When would a 4-wire RTD instead of a 3-wire RTD installation be advisable?

1-8. For best reliability and speed of response, what type of dual element thermocouple should be used?

1-9. What are the performance advantages of a stepped thermowell?

1-10. Why use transmitters rather than the direct wiring of sensors to DCS input cards?

References

1. McMillan, G. K. and Toarmina, C. M. *Advanced Temperature Measurement and Control.* Research Triangle Park: ISA, 1995.

2. Loeffler, R. E. "Thermocouples, Resistance Temperature Detectors, Thermistors - Which?" *Instruments and Control Systems,* May 1973.

3. Kardos, P. W. "Response of Temperature Measuring Elements." *Chemical Engineering,* Aug. 1977.

4. Buckley, P. S. "Dynamics of Temperature Measurements", paper from Process Annual Symposium, Instrumentation Process Industry, Wilmington, Vol. 34, 1979.

5. Stroik, E. R. "RTDs are Sturdier Than You Think." *Instruments & Control Systems,* Jun. 1980.

6. Umrath, E. "Pt RTDs – What are They? Who Needs Them?" *Power,* Jun. 1970.

7. Kerlin, T. W. and Shepard, R. L. *Industrial Temperature Measurements.* Instructional Resource Package (IRP), Student Text. Research Triangle Park: ISA, 1982.

8. Magison, E. C. *Temperature Measurement in Industry.* Research Triangle Park: ISA, 1990.

9. Omega Corp. "Non-Contact Temperature Measurement." *Transactions in Measurement and Control, Vol. 1, 2nd Edition,* Omega Corporation, 1998.

10. Benedict, R. P. *Fundamentals of Temperature, Pressure, and Flow Measurement, 3rd Edition.* John Wiley & Sons, 1984.

11. Trietley, H. L. "Avoiding Error Sources in Platinum Resistance Temperature Detectors." *InTech,* Feb. 1982.

12. Yazbak, G. and Diraddo, R. W. "An Inside Look at Extrusion Melt Temperatures." *Plastics Technology,* Jun. 1993.

13. McMillan, G. K. and Weiner, S. "Straight Talk." *Control Talk, Control,* Jan. 2008.

14. McMillan, G. K. "Is Wireless Process Control Ready for Prime Time?", *Control,* May 2009.

Common Reference Information, Conversions, and Tables may be found at http://www.emersonprocess.com/rosemount/document/tds/3061b00q.pdf

2
Measurement Error

This unit describes the contributors to the accumulated measurement error and some of the design and installation modifications and new sensor technologies to reduce these errors.

Learning Objectives

A. **Understand how installation effects make the accumulated error five times larger than the error limits normally stated for thermocouples (TCs) and resistance temperature detectors (RTDs).**

B. **Recognize the alternatives in terms of sensor insulation, sheath, thermowell, and wiring design and installation to improve the accuracy of TCs and RTDs.**

C. **Appreciate the advantages offered by new measurements such as the Johnson noise thermometer and the optical fiber thermometer.**

2-1. Heat Conduction Error

Thermowells are usually required for process streams because of the need to provide (1) a more rugged and corrosion-resistant barrier between the process and the sensor and (2) a method of removal of the sensor while the process is still in operation. Thermowells are needed to improve the reliability, maintainability, and safety of thermocouple and resistance temperature detector systems.

Unfortunately, thermowells aggravate the measurement error due to nonideal heat transfer. The thermowell increases the diameter and mass of the probe by at least a factor of six and adds an annular air gap around the sensor. The result is a greater heat conduction error and dynamic error.

Since the process fluid is not at the same temperature as the thermowell connection (socket or flange), the sensor tip temperature is between the temperature of the process fluid and the thermowell connection. The temperature difference between the process fluid and sensor tip is the driving force for heat flow between the process and tip. Similarly, the temperature difference between the sensor tip and the thermowell connection is the driving force for heat flow between the tip and the connection. Conservation of energy requires that in the steady-state, the heat flow from the process to the tip must equal the heat flow from the tip to the connection. The result is that these temperature differences will always exist, and their size depends upon the resistance to heat flow in the associated paths.

It is desirable that the thermal resistance between the process and the tip be as small as feasible and the thermal resistance between the tip and the connection be as large as feasible to make the difference between the process and sensor tip, which is called heat conduction error, as small as possible.

The situation is analogous to an electrical circuit, in which the temperature difference, heat flow, thermal resistance, and product of mass and heat capacity are functionally similar to potential difference, current flow, electrical resistance, and capacitance, respectively. The thermal resistance, the key parameter for heat conduction error, is equal to the sum of convective and conductive resistances, where the convective resistance is inversely proportional to the heat transfer coefficient and the conductive resistance is proportional to the material thickness divided by material thermal conductivity. Equations 2-1 and 2-2 show the major thermal resistances at the tip and along the wall of a thermowell, respectively. These equations show the parameters to alter in order minimize conduction error. Equation 2-1 will also be used in Section 2-2 to estimate dynamic error. The convective heat transfer coefficients and thermal conductivities can be found in Appendices D and H, respectively, for common materials. The heat transfer areas in these equations are actually a complex function of the geometry of the thermowell and sensor assembly but can be approximated as the cross-sectional area and computed from thermowell radii per Equations 2-3 and 2-4.

$$R_t = (1/h + x_c/k_c + x_t/k_t) * (1/A_t) \tag{2-1}$$

$$R_w = (x_w/k_w) * (1/A_w) \tag{2-2}$$

$$A_t = 3.14 * r_t^2 \quad (2\text{-}3)$$

$$A_w = 3.14 * (r_o^2 - r_i^2) \quad (2\text{-}4)$$

where:

A_t = heat transfer area at thermowell tip (cross-sectional area based on tip diameter of thermowell) (ft.^{2})

A_w = heat transfer area of thermowell wall (cross-sectional area based on outside and inside diameters of thermowell) (ft.^{2})

r_i = inside radius of thermowell body (ft.)

r_o = outside radius of thermowell body (ft.)

r_t = tip radius of thermowell (ft.)

h = convective heat transfer coefficient between the process fluid and thermowell tip (Btu/hr. * ft.^{2} * °F)

k_c = thermal conductivity of coating (e.g., process or glass) on exterior of thermowell tip (Btu/hr. * ft. * °F)

k_t = thermal conductivity of material of thermowell tip (Btu/hr. * ft. * °F)

k_w = thermal conductivity of material of thermowell wall (probably the same as k_t for tip) (Btu/hr. * ft. * °F)

x_c = thickness of coating (e.g., process or glass) on exterior of thermowell tip (ft.)

x_t = thickness of material of thermowell tip (ft.)

x_w = length of thermowell wall between tip and connection (ft.)

R_t = thermal resistance at the tip between the process fluid and thermowell interior tip (°F*hr/Btu)

R_w = thermal resistance along the wall between the thermowell interior tip and exterior connection (°F*hr/Btu)

To make the sensor tip as sensitive to the process temperature and insensitive to the equipment or pipe wall temperature, the thermal resistance between the tip and the process and between the tip and the pipe or equipment wall should be small and large, respectively.

As stated previously, to reduce the heat conduction error, R_t should be much less than R_w. For normal thermowell geometries, if there is no coating on the thermowell and if the thermowell tip is not ceramic or some other nonthermally conductive material, R_t primarily depends upon the convective heat transfer coefficient, which is generally proportional to the fluid velocity raised to the 0.5 power per equation H-1 in Appendix H. A larger thermowell diameter increases the heat transfer area but the increase in metal mass and thickness and possibly the air gap can dramatically slow down the sensor response increasing the dynamic error (see Section 2-3). R_w depends upon the distance from the tip divided by the thermal conductivity of the thermowell wall. Consequently, to minimize heat conduction error, it is desirable for the convective heat transfer coefficient to be as large, the thermowell wall length be as long, and the thermowell wall thermal conductivity be as small as possible.

To minimize the temperature difference between the thermowell tip and the process, maximize the heat transfer coefficient by increasing the fluid velocity and eliminating process coatings and minimize the mass of the thermowell tip. To maximize the temperature difference between the thermowell tip and the pipe or equipment wall (minimize conduction error), decrease the thermal conductivity and diameter of the thermowell and increase the length of the thermowell.

Numerous articles have developed equations and plots of the heat conduction error as a function of these parameters [1, 2, 3, 4, and 5]. All of the methods have made simplifying assumptions as to the paths and mechanisms of heat transfer. Equations 2-5 through 2-8 from Ref. 1, which assumes a thin homogeneous right circular cylinder attached to a flange at a constant temperature, are useful to estimate the gross conduction error ratio for a temperature probe that is either a thermowell or a bare sheathed sensor. For a polymer with convective heat transfer coefficient of about 10 Btu/hr * ft^2 * °F, the conduction error for various metals is plotted versus a geometric parameter of length divided by the square root of the probe outside diameter in Figure 2-1a for the installation shown in Figure 2-1b. The plot shows that conduction errors become small for stainless and Hastelloy thermowells and sheaths for a geometric parameter greater than 2 $ft.^{0.5}$ for polymer service.

$$C_e = [(1 + p) * e^{-m * L}] / (1 + p * e^{-2m * L}) \quad (2\text{-}5)$$

$$E_1 = C_e * (T_b - T_f) \quad (2\text{-}6)$$

$$m = [(2 * h) / (k * r)]^{0.5} \quad (2\text{-}7)$$

$$p = (k * m - h) / (k * \mathrm{m} + h) \quad (2\text{-}8)$$

where:

C_e = conduction error ratio (dimensionless)

E_1 = temperature error due to conduction (°F)

h = convective heat transfer coefficient (Btu/hr * ft^2 * °F)

k = thermal conductivity of probe wall (Btu/hr * ft * °F)

L = length of probe (ft.)

m = intermediate parameter (dimensionless)

p = intermediate parameter (dimensionless)

r = radius of probe (ft.)

T_b = bulk temperature (°F)

T_f = flange temperature (°F)

In most other applications the convective heat transfer coefficient is larger than that given for polymer service (see Appendix H). Figure 2-2 provides a finer resolution of errors for the more general case of different convective heat transfer coefficients and two installation geometries. As might be expected, the recessed well has conduction errors that are one to two orders of magnitude larger than the fully inserted well.

A seemingly insignificant air gap between the sensor sheath and the inside wall and bottom of the thermowell can greatly increase the thermal resistance between the sensor and the thermowell tip. the result is a larger temperature difference between the sensing element and the tip of the thermowell besides a larger dynamic error (see Section 2-3 for details).

To minimize the temperature difference between the thermowell tip and the sensing element tip, ensure the sheath is seated firmly and tightly in the thermowell tip.

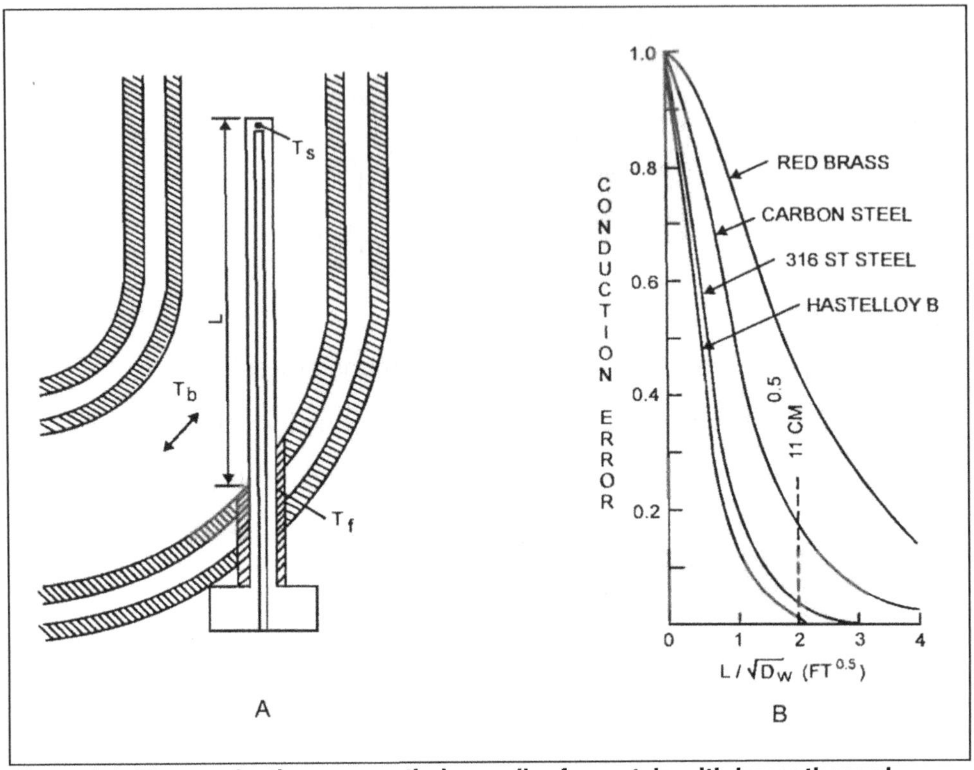

Figure 2-1. The conduction error ratio is smaller for metals with lower thermal conductivities as seen for polymer service where h = 10 Btu/hr*ft* °F *(Source: Ref. 1, Figures 1 and 2. Printed with permission)*

The analysis to this point has not considered the effect of a nonuniform process temperature. For side entry of thermowells halfway into pipelines, parabolic and linear temperature profiles cause temperatures to often read 10% and 20% lower, respectively, due to conduction error [4]. The immersion in the peak temperature region can be increased, and this error for pipe cross-sectional temperature profile can be decreased by installation of the temperature probe at an angle or in an elbow. Also, bare sheathed elements can be bent 90° and turned downstream or spiraled to increase their insertion length [1] [4]. The effect of temperature profile is negligible for high convective heat transfer coefficients (i.e., $h > 1000$).

From the standpoint of minimizing conduction and profile error, a pipeline elbow installation is greatly preferred because it puts the maximum amount of length in the center of the pipe, which is generally the most representative point of the bulk temperature.

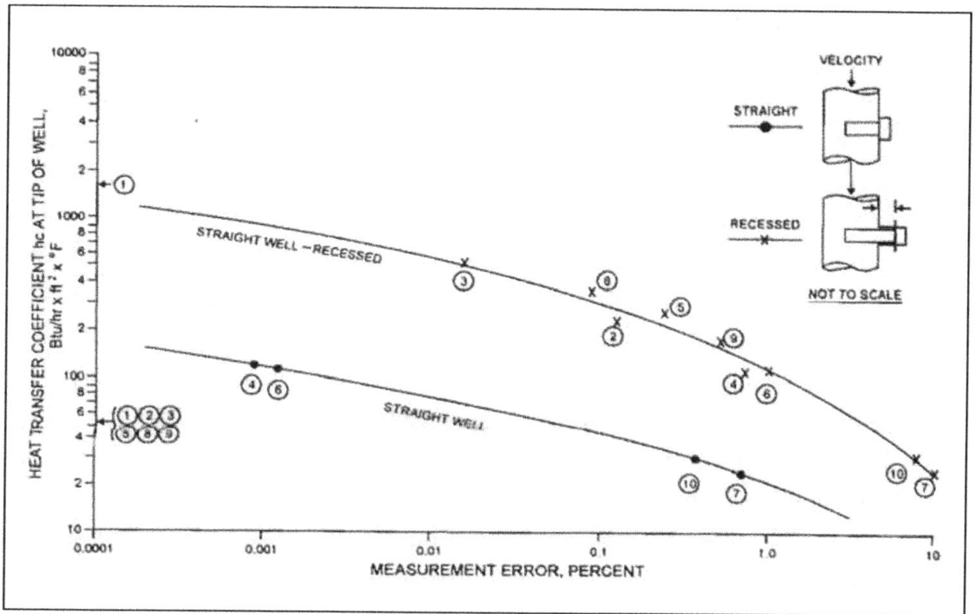

Figure 2-2. The conduction error is larger for recessed thermowells and smaller convective heat transfer coefficients *(Source: Ref. 2, Figure 4. Printed with permission)*

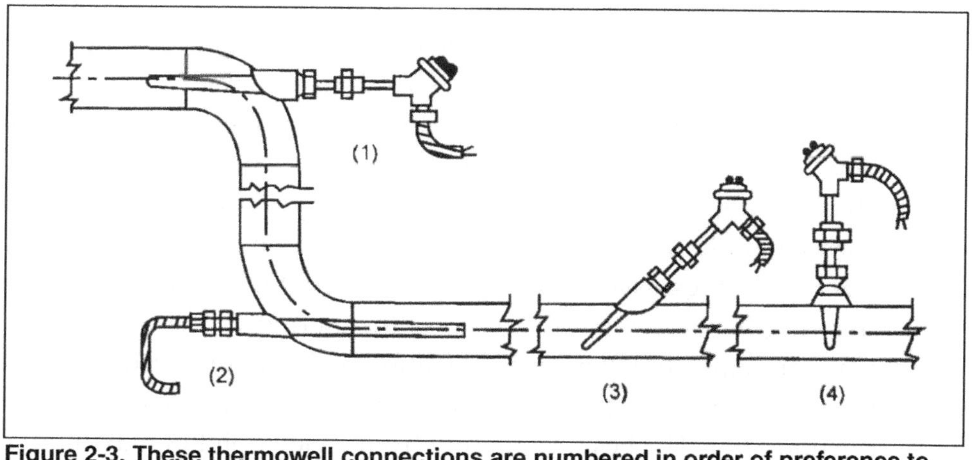

Figure 2-3. These thermowell connections are numbered in order of preference to reduce heat conduction and fluid profile errors *(Source: Masek, J. A., "Thermowell Design for Process Piping," Hydrocarbon Processing, Volume 43, Nos. 2, 3, and 4, 1974, presented by Thermetrics Corporation, Anaheim, CA. Printed with permission)*

The design features that make a sensor fast also make a sensor more accurate by minimizing temperature differences and dynamic error.

For installation point (1) in Figure 2-3 where the thermowell is pointed into the flow, the flow profile is more uniform, whereas the discontinuity of the elbow will introduce some distortion for installation point (2) where the thermowell is pointed downstream. In some cases the pipe diameter may have to be increased at the elbow to avoid excessive reduction of the cross-sectional area and pressure drop. If an elbow or tee is not available, an angle side entry, as shown at point (3) in Figure 2-3, is the next best choice. The perpendicular side entry shown at point (4) is the least desirable of the installations due to short total immersion length (especially for small diameter pipelines) and inadequate center exposure. It has been shown that with side entry thermowells per point (4) into two-inch Therminol™-jacketed polymer pipelines, the sensor responds more to therminol temperature than to process temperature. Figure 2-4 shows how side entry probes can be bent for small connections or spiral wound for large connections to increase their insertion length, particularly in the center line. The support (thin beam or tube) should have a low thermal conductivity and small cross-sectional area to minimize heat conduction along its walls.

As the insertion length is increased, the possibility is increased that the thermowell or bare element will vibrate in resonance with the vortex shedding frequency. Such vibration can cause fatigue failure of the thermowell or sheath welds and a potentially hazardous release of process fluid. Thus, there is an optimum length that minimizes conduction error and vibration. Refs. 3 and 4 provide equations for predicting the maximum allowable thermowell length. Figure 2-5 from Ref. 2 shows how this length varies with fluid velocity for various installation geometries. The fully inserted and recessed thermowells have about the same maximum insertion length. Narrowing of the end of the thermowell results in a longer maximum insertion length by elevation of the natural frequency. Thus, a tapered thermowell permits a longer insertion length, which reduces conduction error. The best installation from a viewpoint of maximum reliability and minimum conduction error is a tapered thermowell installed in a piping elbow.

Table 2-1 from Ref. 6 shows that the maximum fluid velocities for a heavy duty one-inch NPT threaded 316ss thermowell with a 1/4-inch sensor bore are three times larger for gas than for liquid service. These velocities are quite large until insertion lengths greater than ten inches are reached. Such long lengths might be encountered for installation in equipment rather than in piping, but the velocities in these larger areas are generally low. For example, the average bulk velocity in a well-mixed reactor rarely exceeds one foot per second except near the impeller or dip tube tip. It is

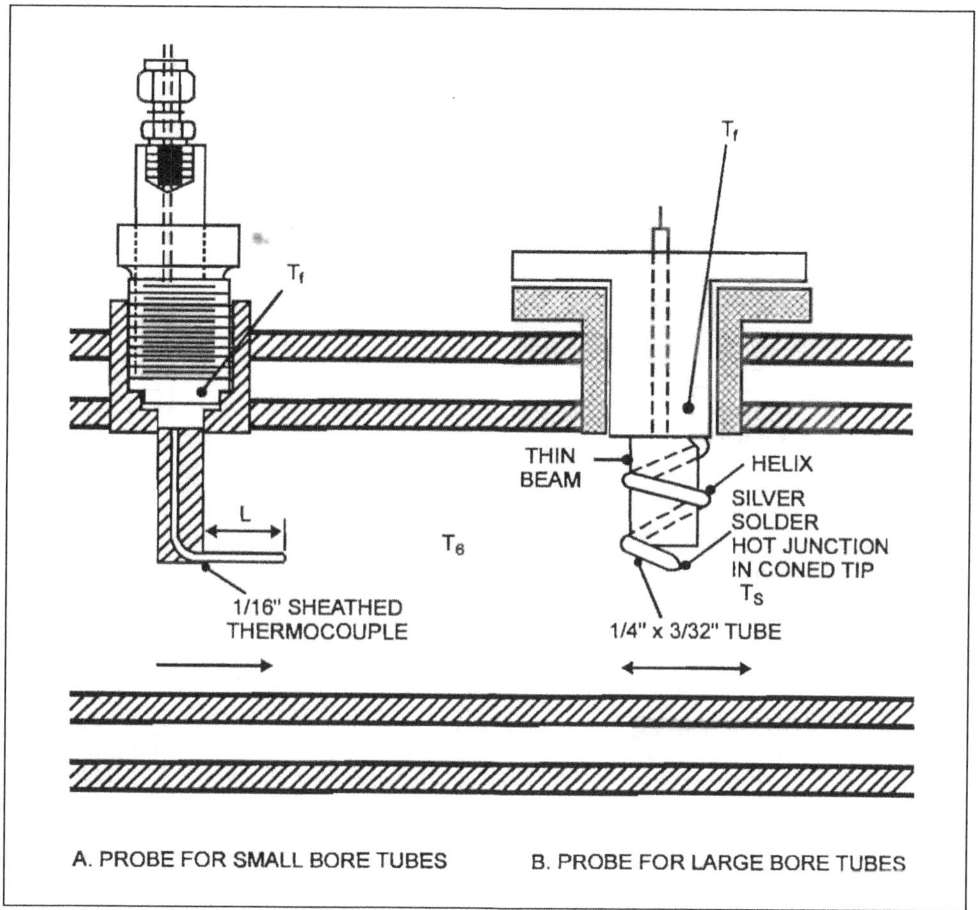

Figure 2-4. These bent and spiral-wound sheathed probes improve the accuracy for small pipelines with side entry installations, particularly for polymer flow *(Source: Ref. 1, Figures 3 and 4. Printed with permission)*

important to note that such installations need longer insertion lengths due to the smaller convective heat transfer coefficients associated with the lower velocities. Often the thermowell diameter is increased to support the longer length. Thermowells with larger outside diameters will naturally have lower maximum velocities unless tapered.

Vibration dampers have been designed for fast, high-velocity streams in exchangers and power plant turbine exhaust systems. A sheathed RTD is encapsulated in high temperature silicone rubber and inserted through an open-ended support well with the sheathed tip exposed to the gas stream. The rubber prevents flexing and cold work failure of the lead wires at the ceramic seal [7].

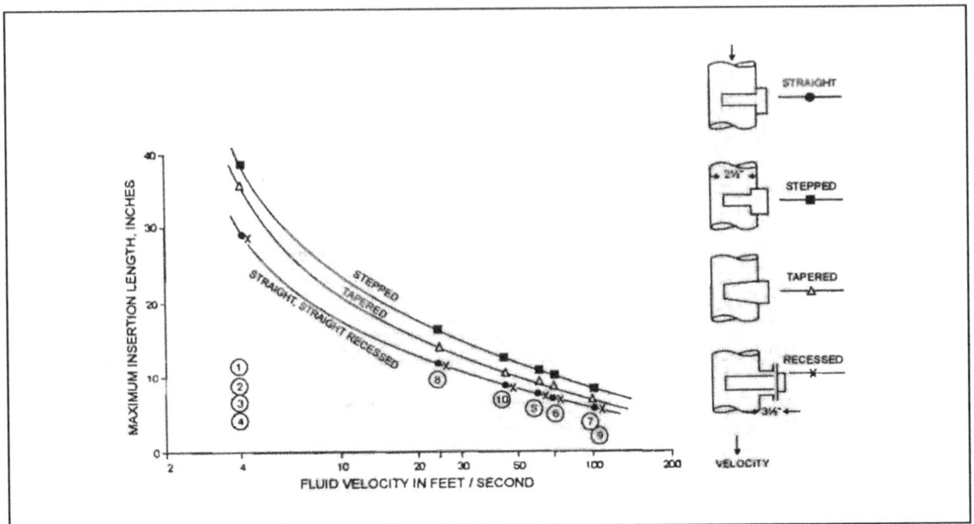

Figure 2-5. The maximum insertion length is larger for uniform narrowing of the sensign end of the thermowell *(Source: Ref. 2, Figure 3)*

Table 2-1. The maximum fluid velocities — three times larger for gas than for liquid flow and quite large for a heavy duty one-inch threaded thermowell *(Source: Ref. 6, Table 2)*

Insertion Length (Inches)	Gas Velocity (ft/sec)	Liquid Velocity (ft/sec)
2.5	490	163
3.0	309	103
4.5	228	76
6.0	156	52
7.5	83	27
10.5	42	14
13.5	26	8
16.0	17	5
19.5	12	4
22.5	9	3

2-2. Radiation Error

Radiation error can develop in the presence of a flame or when the surrounding piping or equipment surface temperature is significantly different from the exterior probe temperature. For combustion zones the emissivities and radiation paths between the flame, equipment surfaces, and thermowell must be considered. In these cases, the calculations become too complex, and radiation corrective installation designs should be used as a matter of course where conditions exist for the radiation error. It doesn't typically occur in liquid service because the transmittance is too low for appreciable energy transfer by radiation. For gas service, where the gas transmittance and emissivity are negligible, the error in temperature reading can be approximated by Equation 2-9 for radiation heat transfer from a temperature probe to a cooler duct surface. The equation shows that the radiation error can be reduced by polishing the surface to reduce its emissivity, increasing the velocity past the probe to enlarge the convective heat transfer coefficient, increasing the length or decreasing the diameter of the probe, and adding intermediate surfaces called radiation shields to reduce the temperature difference between the probe and viewpoint surfaces. Each shield assumes a temperature between those of its viewpoint surfaces so that the layering of concentric shields successively reduces the radiation error. Alternatively, a shield can be heated and its temperature measured and controlled to be close to the probe temperature. A simple high velocity thermocouple assembly uses a long slender thermocouple support, a baffled high velocity gas stream, and a radiation shield to reduce the temperature measurement error. The gas stream velocity is incremented until the increase in the indicated temperature is negligible [8].

Concentric shields can reduce the radiation error from heat radiated from the thermowell, protection tube, or sensor tip to surrounding surfaces.

$$E_2 = [(G * E * r) / (h * L)] * (T_o^4 - T_d^4) \qquad (2\text{-}9)$$

where:

E	=	emissivity of the surface of temperature probe
E_2	=	temperature error due to radiation (°R)
G	=	Stefan-Boltzman constant (0.1783×10^{-8} Btu/hr * ft^2 * °R^4)
h	=	convective heat transfer coefficient (Btu/hr * ft^2 * °R)
L	=	length of temperature probe (ft.)
r	=	radius of temperature probe (ft.)
T_o	=	temperature of outside surface of probe (°R)
T_d	=	temperature of duct interior surface (°R)

A multiple-shield high velocity assembly with a bare thermocouple can make the radiation error negligible, but is not as reliable due to corrosion and decalibration of the exposed thermocouple junction and pluggage of the small passageways particulates and tar. The first five measurement errors in Table 2-2 are from tests documented in Ref. 9 for various types of assemblies for a true gas temperature of 2105°F and a wall temperature of 311°F. The last measurement error is based on a computation in Ref. 8 for a thermowell made of 1/4-inch schedule 40 steel pipe that extends one foot into the gas stream. The error for thermowell consists of a radiation error of 1150°F and a conduction error of 50°F.

Table 2-2. The measurement error due to radiation and conduction for the extreme case of a true temperature of 2105 and a wall temperature of 311 degrees Fahrenheit is large except for a multiple shield high velocity thermocouple assembly *(Source: Ref. 9. Printed with permission)*

Thermocouple Assembly Description	Measurement Error
Multiple-shield high velocity thermocouple	2°F
Bare 28-gauge thermocouple	110°F
Simple high velocity thermocouple	142°F
Bare 14-gauge thermocouple	183°F
Bare 8-gauge thermocouple	242°F
Thermowell 1/4-inch schedule 40 one foot long	1200°F

2-3. Dynamic Error

The dynamic error originates from the time constant, which is caused by thermal resistance to heat transfer in the probe, and can be approximated per Equation 2-10 as the rate of change of temperature multiplied by the measurement time constant. The response of a temperature sensor in a thermowell actually consists of two time constants as shown in Table 2-3 [10]. For most applications, these can be summed and the total used for the measurement time constant in Equation 2-10 [11].

$$E_3 = (dT/dt) * \tau_m \quad (2\text{-}10)$$

where:

E_3 = temperature error due to dynamics (°F)

dT/dt = rate of change of temperature (°F/min)

τ_m = measurement time constant (min)

A comparison of Table 2-3 for thermowell assemblies and Table 2-4 for bare sensor elements shows that most of the measurement lag of typical sensor assemblies is due to the poor thermal conductivity and, thus, high thermal resistance of the annular air gap in the thermowell. This resistance can be reduced by the addition of thermally conductive oil or grease in the tip of the thermowell. The fill must not extend much beyond the tip or it will increase the conduction error. Also, the temperature must not exceed the degradation or boiling point of the fill. Fine metal shavings and particles can help if they lie in the annular gap near the tip and do not accumulate in the end of thermowell.

Table 2-3. The measurement time constant for a thermowell is the summation of two time constants, where the magnitude of the largest depends heavily upon the size and fill of the annular clearance near the tip

Fluid Type*	Fluid Velocity, fps	Annular Clearance, inch	Annular Fill	Time Constants, seconds
Gas	5	0.04	Air	107 and 49
Gas	50	0.04	Air	93 and 14
Gas	152	0.04	Air	92 and 8
Gas	300	0.04	Air	91 and 5
Gas	152	0.04	Oil	22 and 7
Gas	152	0.04	Mercury	17 and 8
Gas	152	0.02	Air	52 and 9
Gas	152	0.005	Air	17 and 8
Liquid	0.01	0.01	Air	62 and 17
Liquid	0.1	0.01	Air	32 and 10
Liquid	1	0.01	Air	26 and 4
Liquid	10	0.01	Air	25 and 2
Liquid	10	0.01	Oil	7 and 2
Liquid	10	0.01	Mercury	2 and 0.2
Liquid	10	0.055	Air	228 and 1
Liquid	10	0.005	Air	4 and 1

* The gas is saturated steam and the liquid is organic.

The measurement time constant for a temperature probe can be estimated as the thermal capacitance (mass multiplied by heat capacity of tip) multiplied by the sum of the thermal resistance associated with the thermowell tip (R_t) and the sensor (R_s), expressed by Equations 2-1 and 2-11, respectively [12]. These equations coupled with Equation H-1 in Appendix H and the physical property data in Appendix D are useful for showing the relative effects of velocity, coating, thermowell material, and air gap size on the measurement time constant. For example, a decrease in the thermowell material conductivity or the formation of a thick coating increases the resistance to heat transfer by convection per Equation H-1 as well as the resistance to heat transfer by conduction per Equation 2-11. The result is a double effect. Thus, the use of Hastelloy B, which has a thermal conductivity of 72, instead of 316ss, which has a thermal conductivity of 113, for a thermowell, causes the measurement time constant to increase from 19 to 93 seconds in water and from 37 to 125 seconds in light oil [12].

To minimize the dynamic error in liquid streams, the measurement time constant must be as small as possible by selecting a location with a velocity large enough to keep the thermowell clean and responsive and a design that decreases the mass of the metal at the tip and the air gap in the thermowell.

Table 2-4. The measurement time constant for a bare sensor is much less than that for a thermowell and is slightly larger for a resistance temperature detector (RTD) than for a thermocouple

Bare Sensing Element Type	Time Constant (seconds)
Thermocouple 1/8-inch sheathed and grounded	0.3
Thermocouple 1/4-inch sheathed and grounded	1.7
Thermocouple 1/4-inch sheathed and insulated	4.5
Single Element RTD 1/8 inch	1.2
Single Element RTD 1/4 inch	5.5
Dual Element RTD 1/4 inch	8.0

For a still fluid, the thermal lag from poor convection becomes large for liquid and tremendous for gas service. For example, the measurement time constant for still water increases by a factor of four and, for still air, increases by a factor of 80 compared to water at 15 feet per second [12].

$$R_s = (x_f/k_f + x_s/k_s + x_i/k_i) * 1/A_s \tag{2-11}$$

$$\tau_m = (M * C) * (R_t + R_s) \tag{2-12}$$

where:

A_s = heat transfer area of sensor tip (cross-sectional area based on outside diameter of sensor) (ft.^{2})

C = heat capacity of tip (Btu/lb * °F)

k_f = thermal conductivity of fill (e.g., air or oil) of interior of thermowell tip (Btu/hr * ft * °F)

k_s = thermal conductivity of sensor sheath (e.g., stainless steel) in thermowell tip (Btu/hr * ft * °F)

k_i = thermal conductivity of sensor insulation (e.g., magnesium oxide) in thermowell tip (Btu/hr * ft * °F)

M = mass of tip (lb)

R_s = thermal resistance between interior tip of thermowell and sensor (e.g., thermocouple junction) (°F * hr/Btu)

R_t = thermal resistance at the tip between the process fluid and interior tip of the thermowell (°F * hr/Btu)

To recapitulate at this point, an increase in the thermal resistance of the tip (R_t) tends to increase the conduction, radiation, and dynamic error, whereas an increase in the sensor thermal resistance (R_s) primarily affects the dynamic error.

Except for thermowell thermal conductivity, the design and installation considerations that minimize dynamic error also minimize conduction error.

2-4. Velocity Error

If the temperature probe was moving at the same velocity as a gas stream, it would measure the static temperature of the gas. Since the probe is stationary, there is a temperature rise from compression heating as a result of the gas being brought to rest around the temperature probe. The total temperature is equal to the static temperature plus this temperature rise. The probe actually sees a recovery temperature that lies between the static and the total temperature. It is approximated by multiplication of the temperature rise by a recovery factor (r). This recovery factor is about 0.65 for flow perpendicular and 0.9 for flow parallel to the axis the probe. Vortex thermometers have been designed to give a recovery factor of zero so that the thermometer measures the static temperature, but they are complex and their performance is unpredictable [13].

If a measurement of static temperature is desired, the velocity error can be estimated by Equation 2-13, where the static temperature is in absolute degrees Rankine. The error increases with the square of the Mach number.

For example, if a perpendicular air stream's Mach number increases from 0.2 to 0.6 with a static temperature of 620°F, the temperature rise increases from 5.6 to 50.3°F ($C_p/C_y = 1.4$ for air). If a measurement of total temperature is desired, the velocity error can be estimated by Equation 2-14 where the factor (r - 1.0) has been substituted for r. Note that the error in this case is negative since the sensed temperature is less than the total temperature.

If a measurement of static temperature is desired:

$$E_4 = 0.5 * r * (C_p/C_v + 1) * M^2 * T_g \qquad (2\text{-}13)$$

If a measurement of total temperature is desired:

$$E_4 = 0.5 * (r - 1.0)\,(C_p/C_v + 1) * M^2 * T_g \qquad (2\text{-}14)$$

where:

C_p = specific heat of gas at constant pressure (Btu/lb * °F)

C_v = specific heat of gas at constant volume (Btu/lb * °F)

E_4 = temperature error due to velocity effect (°F)

M = Mach number (dimensionless)

r = recovery factor (dimensionless)

T_g = static temperature of gas (°R)

For high velocity gas streams, the velocity error can be estimated for a given velocity if the fluid specific heats are known.

2-5. Electronic Error

The electronic error is rather simple to estimate per Equation 2-15 compared to the previous errors. It consists primarily of a circuit error expressed as a percent of span. The error ranges from 0.05% to 0.25% of span for transmitters and from 0.3% to 0.9% of span at reference conditions for thermocouple and RTD input cards to a distributed control system (DCS), as shown in Tables 2-5 and 2-6. These cards often have input ranges of several hundred to several thousand degrees Fahrenheit. Furthermore, the operating error is often about twice as large. The result is excessive measurement error. For example, a type S thermocouple computer input card per Table 2-6 has an operating error of about 24°F. The operating temperature effect on field transmitters from a performance report of tests ranged from 0.1% to 0.3% per 10°F change in ambient temperature. The error is much less for smart transmitters that compensate operating temperature.

Table 2-5. The thermocouple input card operating error is about twice the reference error *(Source: Fisher-Rosemount bulletin 4.2:CL6821)*

Input Range	Accuracy ref.	Accuracy Oper.	Thermocouple Type	Sig. Cond. Module Type
–148 to 1400°F (–100 to 760°C)	0.31%	0.61%	Type J High	CL6854-1
–60 to 640°F (–51 to 338°C)	0.33%	0.63%	Type J Low	CL6854-2
–148 to 2462°F (–100 to 1350°C)	0.31%	0.63%	Type K High	CL6854-3
0 to 1000°F (–18 to 538°C)	0.33%	0.63%	Type K Low	CL6854-4
32 to 3182°F (0 to 1750°C)	0.35%	0.65%	Type R	CL6854-5
–300 to 600°F (–184 to 315°C)	0.35%	0.65%	Type T	CL6854-6
–100 to 1600°F (–73 to 871°C)	0.30%	0.60%	Type E	CL6854-7
32 to 3272°F (0 to 1800°C)	0.45%	0.75%	Type B	CL6854-8
23 to 3182°F (0 to 1750°C)	0.35%	0.65%	Type S	CL6854-9

For digital devices, there is also a slight analog-to-digital (A/D) conversion error that is the span divided by the resolution of the convertor determined by the number of bits. While A/D effect on measurement error is often negligible, as seen in Equation 2-15, its effect on the control error is often significant because the staircasing that occurs for slowly changing temperatures severely restricts the use of derivative action, a controller mode that is otherwise extremely beneficial for temperature loop performance. The A/D noise effect on controller output, as estimated by Equation 2-16, increases with the span of the computer input card and the derivative and gain settings of the controller. For transmitters, the measurement span is typically the same as the span of the controller scale, so that the effect of span cancels out. The wide spans of thermocouple and RTD computer input cards are disastrous for temperature loops in general and reactor loops in particular, since process dynamics permit gain and derivative settings of ten or more.

Table 2-6. The RTD input card operating error is several times larger than the sensor error *(Source: Fisher-Rosemount bulletin 4.2:CL6821)*

Input Range	Accuracy ref.	Accuracy Oper.	Sig. Cond. Module Type
–270 to 140°F (–168 to 60°C)	0.40%	0.70%	CL6853-1*
–148 to 212°F (–100 to 100°C)	0.40%	0.70%	CL6853-2*
–100 to 600°F (–73 to 315°C)	0.39%	0.69%	CL6853-3*
32 to 382°F (0 to 200°C)	0.42%	0.72%	CL6853-4*
32 to 1112°F (0 to 600°C)	0.36%	0.65%	CL6853-5*
100 to 500°F (38 to 280°C)	0.43%	0.73%	CL6853-6*
–270 to 140°F (–168 to 60°C)	0.40%	0.70%	CL6853-7**
–148 to 212°F (–100 to 100°C)	0.40%	0.70%	CL6853-8**
–100 to 600°F (–73 to 315°C)	0.39%	0.69%	CL6853-9**
32 to 392°F (0 to 200°C)	0.42%	0.72%	CL6853-10**
32 to 1112°F (0 to 600°C)	0.35%	0.65%	CL6853-11**
100 to 500°F (38 to 260°C)	0.48%	0.73%	CL6853-12**
32 to 300°F (0 to 149°C)	0.90%	1.20%	CL6853-13***

* 100 ohm platinum with temperature coefficient of 0.00382 ohm/°C
** 100 ohm platinum with temperature coefficient of 0.00385 ohm/°C
*** 10 ohm copper

For a type J high thermocouple card, a controller scale span of 100°, a twelve-bit A/D, a gain setting of 10, a derivative setting of 0.8 minute, and a derivative filter of 0.1 minute, the controller output noise would be about 30%. Most industrial distributed control systems (DCS) have a filter built into the controller algorithm that is one-eighth of the derivative time setting. For these systems, the ratio of derivative to filter time (T_d/T_f) in Equation 2-16 is eight, and Equation 2-16 simplifies to Equation 2-17. Interestingly enough, the noise from the derivative mode becomes independent of rate setting. On the other hand, the change in controller output from the proportional mode reaction to the A/D step is spread out by the filter and actually becomes more abrupt as the rate setting is decreased due to the corresponding reduction in the filter time. The response from the integral mode (reset action) to the A/D step increases with filter time setting but is negligible compared to the effect of the other modes for properly tuned loops.

$$E_5 = S_e * (E_e + 1/2^n) \qquad (2\text{-}15)$$

$$O_c = (100\%/2^n) * (S_e/S_c) * K_c * (T_d/T_f + 1) \qquad (2\text{-}16)$$

For most DCS with $T_d > 0$:

$$O_c = (100\%/2^n) * (S_e/S_c) * K_c * 9 \qquad (2\text{-}17)$$

where:

E_e = electronic error (fraction of span)

E_5 = electronic error (°F)

K_c = controller gain (dimensionless)

n = number of microprocessor bits less sign bit ($n = 11$ for 12 bit A/D)

S_e = electronic span (°F)

S_c = controller span (°F)

T_d = derivative time (rate) setting (min)

T_f = filter time setting (min)

To summarize, the electronics error for temperature transmitters is generally much less than for DCS input cards due to the narrower span and better accuracy specifications of transmitters. Also, the noise from the controller's derivative and proportional mode reaction to the larger A/D convertor step from input cards is detrimental. For these reasons, temperature loops should use transmitters instead of TC or RTD computer input cards.

The electronic error is excessive for RTD and TC input cards particularly for the older design DCS. All temperature loops should use transmitters.

2-6. Sensor Error

The sensor error is what is most often used to express the measurement error, even though from this unit it is obvious that many more types of errors can contribute to the measurement error. ISA tolerances provide an indication of the initial calibration variations of TCs and RTDs from production batch to production batch. Within a particular batch, the variations can be expected to be less. For example, tests made on type K thermocouples at 1800°F showed deviations of less than +9°, whereas the ISA tolerance is + 13.5° [14]. The variations can largely be compensated for by calibration adjustments within a transmitter. This is, of course, not feasible for a TC or RTD input card to computers because of the many different sensors connected to the card. Thus, for computer inputs, the tolerance can be used as a conservative estimate of the sensor error E_6.

The ISA tolerances are shown in Table 2-7 for the valid ranges of TCs and RTDs. The table shows that RTDs have tolerances much smaller than TCs below 400°F. Above this temperature, the tolerances for type R and S thermocouples start to become less than those for RTDs [15]. It is important to recognize that this is only part of the story in that extension wire and decalibration errors of TCs to be discussed in succeeding sections make the TC measurement error larger.

The error from sensor tolerance for A grade is about half error for B grade. Sensor matching calibration of the transmitter can eliminate most this error.

The RTD has the best repeatability of industrial temperature sensors. Its repeatability is one to two orders of magnitude better than that of a thermocouple or a thermistor per Table 2-8 [16]. The stability of RTDs, at the high end of their temperature range, deteriorates but still is better than that for thermocouples. Tests by a manufacturer of RTDs calibrated at 932° showed that the reading increased by 0.28° at 500° and 1.03° at 700°F over a period of 6800 hours. Similar tests in this temperature range of thermocouples showed drifts several times larger.

Table 2-7. The ISA tolerances for type R & S thermocouples are slightly better than those for PRTDs above 400° and PRTDs are not suitable for temperatures above 1200°F [15]

Sensor Temp., °F	PRTD E_6, °F		Type T E_6, °F		Type J, K, & N E_6, °F		Type R & S E_6, °F	
	Grade		Grade		Grade		Grade	
	A	B	A	B	A	B	A	B
–300	0.8	2.0	2.9	5.4	—	—	—	—
–150	0.5	1.2	1.4	2.7	—	—	—	—
0	0.2	0.5	0.9	1.8	2.0	4.0	1.1	2.7
200	0.5	1.2	0.9	1.8	2.0	4.0	1.1	2.7
400	0.8	2.0	1.4	2.7	2.0	4.0	1.1	2.7
600	1.2	2.7	2.7	4.1	2.2	4.1	1.1	2.7
750	1.5	3.4	2.9	5.4	2.9	5.4	1.1	2.7
900	1.8	4.3	—	—	3.6	6.8	1.1	2.7
1100	2.0	5.1	—	—	4.3	8.1	1.1	2.7
1200	2.2	5.4	—	—	4.7	8.8	1.1	2.9
1300	—	—	—	—	5.0	8.8	1.3	3.2
1500	—	—	—	—	5.8	11.0	1.4	3.6
1650	—	—	—	—	6.5	12.0	1.6	4.1
1800	—	—	—	—	7.2	13.5	1.8	4.5
2000	—	—	—	—	7.9	14.9	2.0	5.0
2200	—	—	—	—	8.6	16.2	2.2	5.4
2400	—	—	—	—	9.4	17.6	2.4	5.9

Note: Grade A is special, Grade B is standard, and PRTD denotes an industrial platinum resistance temperature detector.

Table 2-8. The repeatability of the RTD greatly exceeds that for either the thermocouple or thermistor [16]

Criteria	Thermocouple	PRTD	Thermistor
Repeatability (°F)	0.2 - 2	0.05 - 1	0.2 - 2
Drift (°F/yr)	0.2 - 2	0.02 - 0.2	0.02 - 0.2
Sensitivity (°F)	0.1	0.002	0.0002
Temperature Range (°F)	–300 - 3000	–200 - 1600	–150 - 550
Signal Output (volts)	0 - 0.06	1 - 6	1 - 3
Power (watts at 100 ohm)	1.6×10^{-7}	4×10^{-2}	8.1×10^{-1}
Minimum Diameter (inches)	0.015	0.125	0.014
Linearity	Good	Excellent	Poor

The construction of the RTD determines its operating range. Low temperature RTDs with low temperature leads perform well between -330°F and +480°F. High temperature RTDs extend the high temperature limit to about 1200°F. To get to the upper limit of 1600°F, a special platinum resistance thermometer (SPRT) is needed that uses a higher purity platinum and has a special construction, a special calibration, and a special price. The cost of an industrial platinum RTD is about $75, whereas the cost of an SPRT escalates to $3000 for the element and $2000 for the calibration [15].

There are several different standards for RTDs with different ohm versus temperature relationships. The most commonly used standard is DIN 43760, which has been adopted by the International Electrotechnical Commission (IEC) and the American Society for Testing and Materials (ASTME). It has a temperature coefficient of 0.21388 ohm per degree Fahrenheit (0.385 ohm per degree centigrade). However, the standard issued by the Scientific Apparatus Makers Association (SAMA) that has a coefficient of 0.21677 ohm per degree Fahrenheit (0.3902 ohm per degree centigrade) is sometimes used. The temperature error for the calibration of a transmitter with the wrong standard can be estimated by Equation 2-18. At 212°F, the error from connecting a DIN standard sensor to a transmitter calibrated for a SAMA standard sensor, is about 2.3°F. However, the unsupported standard issued by the former Scientific Apparatus Makers Association (SAMA) is still sometimes used.

$$E_6 = (T_s - 32) * [(dR/dT)_c - (dR/dT)_s] / (dR/dT)_c \tag{2-18}$$

where:

E_6 = temperature error from RTD standard mismatch (°F)

$(dR/dT)_c$ = temperature coefficient of RTD standard used for calibration of circuit (ohm/°F)

$(dR/dT)_s$ = temperature coefficient of RTD standard for sensor (ohm/°F)

T_s = temperature of RTD (°F)

The thermistor has the best temperature sensitivity but poorest linearity [16]. The resistance of a thermistor decreases by about a factor of two for an increase in temperature of 40°F. The sensitivity of the RTD, while not as good as a thermistor, probably exceeds most application requirements. The RTD has a temperature sensitivity better than 0.002°, but the TC sensitivity seldom goes beyond 0.1°F [15].

The thermocouple output signal is at a much lower level and is more susceptible to electrical interference. There is also a Ettingshausen-Nernst effect, which is the thermomagnetic analog of the Hall effect. Errors of 150% at 212°F have been observed for heater sheath type K thermocouples. The errors were negligible above 300°F [14].

Sheathed thermocouples are faster than comparable sheathed RTDs. However, the difference is a matter of a few seconds, and once they are installed in a thermowell, the annular air gap, for the most part, determines the speed of response of the probe. Thermocouples are also more vibration resistant, but the vibration analysis for the thermowell should dictate the insertion lengths necessary for reliability. Finally, thermocouples are less expensive, but the incremental cost of about $60 should not be a consideration for a temperature control loop important for quality, conversion, capacity, or yield.

The main advantage of thermocouples is the higher upper limit of operating temperature. Below 1200°F, the lower accumulated error, and better stability and sensitivity of RTDs is preferable for important temperature loops.

The emittance adjustment of radiation pyrometers must be set equal to the product of the target's emittance and the window's transmittance. The target's emittance depends upon the emissivity of the material (see Appendix E) and the geometric shape of the object. The window's transmittance depends upon the type and thickness of the transparent medium. The errors caused by various windows at 1400°F for Ircon type R infrared pyrometers are shown in Figure 2-6. The errors from fused quartz glass and polystyrene are negligible.

Two wavelength or ratio pyrometers detect radiation at two different wavelengths and eliminate emittance uncertainties by computing temperature from the ratio of the two signals. However, the target's emittance varies with wavelength so that the cancellation of the emittance effect is incomplete. They are particularly useful for making more accurate measurements of targets that do not fill the sensor's field of view or that are partially obscured by cool dust (e.g., cement kilns) [18].

Another method to reduce emittance uncertainty is to shape the target so that it behaves more like an ideal emitter or blackbody (i.e., unity emittance). This can be approached by making the target concave (e.g., drilling a hole and sighting on the hole) [18]. Table 2-9 gives the approximate error

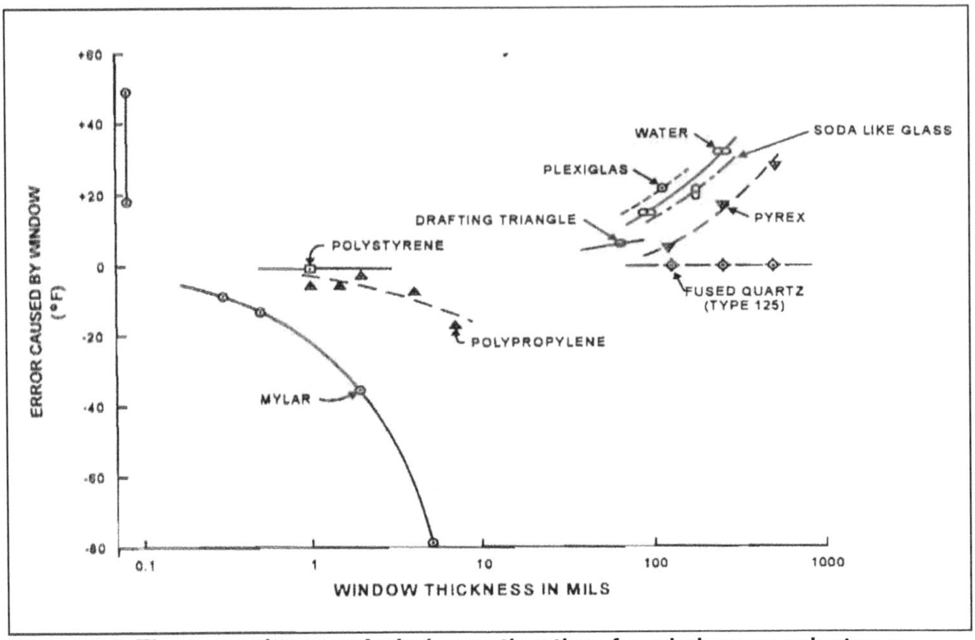

Figure 2-6. The transmittance of windows other than fused glass or polystyrene cause a large error for infrared pyrometers [16]

for various target emittances at 0.65 microns for a disappearing filament calibrated for blackbody conditions [17]. Lasers have been developed to measure the emissivity of the target.

Table 2-9. A correction must be added to the temperature reading to get the true temperature of a target at different emittances for a pyrometer calibrated for blackbody conditions [16]

Target Emittance	Correction at 1652°F	Correction at 1832°F	Correction at 2012°F
0.10	295	351	414
0.25	167	200	234
0.40	108	148	149
0.50	81	95	112
0.60	59	70	81
0.80	25	31	34
0.90	13	14	16
1.00	0	0	0

For a known emittance, the accuracy ranges from 0.3% to 3% of full scale at reference conditions for infrared pyrometers. The most common accuracy specification is 0.75% of full scale.

The interference from other sources such as flames can be eliminated by extension of the sighting tube. A clean nonabsorbing gas purge should be used to reduce the need to clean the window of the tube and the exposed lens of the thermometer [19].

Concave target surfaces, sighting tubes, gas purges, and fused quartz windows maximize optical pyrometer accuracy.

Radiation pyrometers are rarely as accurate as thermocouples and are mostly used for areas that are inaccessible, moving, or beyond the upper range limit (3200°F) of thermocouples.

2-7. Nonlinearity Error

The nonlinearity of thermocouples varies with type and range as exemplified by the deviation of the coefficient (millivolts per 100°) curves in Figure 2-7. Types T and E show the greatest variation of coefficient. The error in percent of full scale at mid scale, when both end points are calibrated correctly, can be estimated by Equation 2-19. For example, the error for a range of 0 to 500°F is -5% (i.e., 25°) for type T and -1.7% (i.e., 8.5°) for type J thermocouples [19]. Over a range of 0°F to 900°, the type T error is 13% and the type J error is 4% compared to 2% for an RTD [20].

$$E_7 = (S_e/4) * (K_{min} - K_{max}) / (K_{min} + K_{max}) \tag{2-19}$$

where:

E_7 = temperature error at mid scale due to nonlinearity (°F)

K_{min} = sensor coefficient at low end of range (ohm or mV/°F)

K_{max} = sensor coefficient at high end of range (ohm or mV/°F)

S_e = temperature span of electronics calibration range (°F)

The nonlinearity of temperature sensors is not as much a concern with the advent of the microprocessor, which enabled polynomials for linearization of the signal. For example, the nonlinearity for a type J can be reduced to 0.2°F by a fifth-order polynomial. The fit is good only within the specified range. Table 2-10 shows the former National Bureau of Standards (now the National Institute of Standards and Technology) polynomial order, valid range, and error for the major types of thermocouples [21].

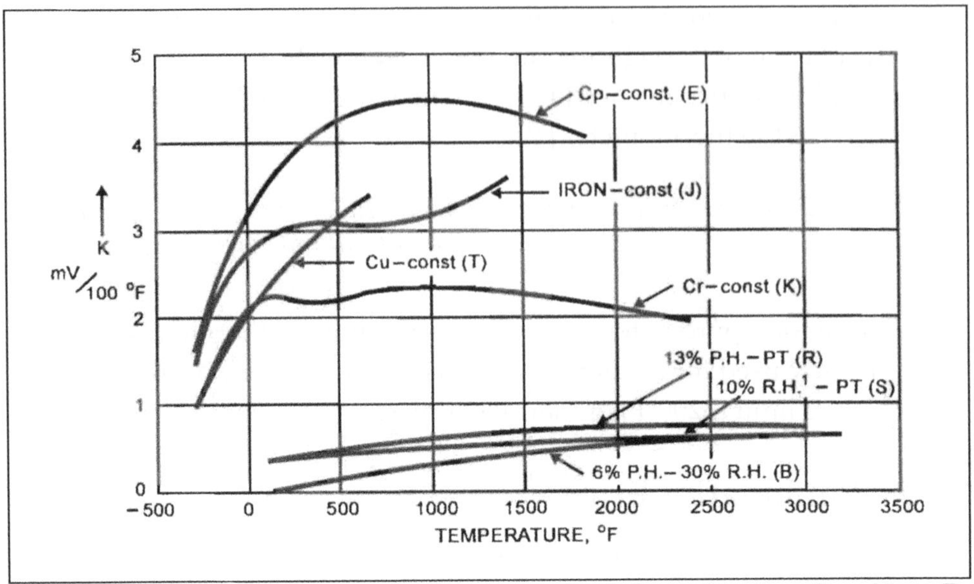

Figure 2-7. The nonlinearity error for thermocouples is largest for type T and type E

Operation outside the range can cause large errors. Also, high-order polynomials tend to have small humps and local gain reversals that are difficult to spot but can cause erratic controller output from the proportional and derivative modes. For loops capable of tight control through high gain action, the deviations from set point might be smaller for an unlinearized signal. For these applications, it is best to omit the linearization and, instead, either narrow the calibration range or bias the set point to reduce the nonlinearity error. The best general solution from a standpoint of the minimization of both measurement and control error is a narrow-range, unlinearized transmitter. A wider-range, linearized measurement is often added as an indicator to show temperatures outside the control band during abnormal operation or start-up and shutdown.

Table 2-10. Polynomials greatly reduce the nonlinearity error within the specified range but create A/D noise in controller [21]

Type	E	J	K	R	S	T
Order	9th	5th	8th	8th	9th	7th
Range (°F)	–148	+32	+32	+32	+32	–256
	1832	1400	2500	1832	3182	752
Error E_7 (°F)	0.9	0.2	1.3	0.9	1.8	0.9

2-8. Decalibration Error

Decalibration of thermocouples occurs for changes in the metallurgical state or composition of the thermoelements that destroy the homogeneity of the thermocouple. The millivolts developed by the thermocouple then depend upon the location of the nonhomogenous portion and temperature gradients. Consequently, recalibration is usually not possible [14]. This can be caused by cold working of the metal from rough handling, pulling of wire through conduit and vibration [21]. The more frequently discussed causes are high temperature and temperature cycles.

Temperature cycling of type K thermocouples between 50°F and 1200°F produces a hysteresis effect and errors of about 1% of reading, caused by an order and disorder transformation where the nickel and chromium atoms occupy specific sites in the chromel lattice at lower temperatures and become randomly distributed at higher temperatures [14].

At temperatures above 1100°, decalibration occurs from composition changes in the thermoelements. Between 1100°F and 1500°F, changes in the Chromel™ element, and above 1800°F, changes in the Alumel™ element are the major sources of the decalibration error for type K. In a test at 2100°F for 50 hours, a stainless steel sheathed type K thermocouple drifted down 24°. The use of an Inconel™ sheath reduced the error to about 2°. The sheath reduces the partial pressure of oxygen inside, which prevents the formation of the protective oxide coating that normally develops at high temperatures. The sheath is also the source of contaminants. The result is higher decalibration errors, especially for small diameter sheathed sensors [14].

Tests of noble metal thermocouples show that they are incompatible with base metal sheaths. Figure 2-8 shows that the drift rate for a type S thermocouple at 2200°F was only 0.01° per minute for a 90% platinum -10% rhodium sheath but increased to 0.3° per minute and 0.5° per minute for stainless and Inconel™ sheaths, respectively [14].

Tests in air for 50 hours at temperatures between 2000°F and 2400°F by a major manufacturer of thermocouples show that with the proper selection of sheath materials, the drift occurred early and was within the ISA tolerance specifications [22].

For thermocouples with proper sheath materials in a noncorrosive or reducing atmosphere:

$$E_8 = E_6 \text{ (wire tolerance)}$$

Types K, E, N, R, and S thermocouples, used for the higher temperature ranges, must be protected from reducing atmospheres.

Excessive loss of chromium from the Chromel™ thermoelement of type K in reducing atmospheres between 1500° and 1900° causes "green rot" and a large negative change in the measurement [22].

For applications above the temperature rating of metal sheaths, with gaseous contaminants or reducing conditions, primary (outer) and secondary (inner) ceramic protection tubes are used to prevent composition changes in the thermoelements [23]. Ceramic materials are chosen to allow only the beneficial diffusion of hydrogen. The dynamic error is greater for sensors with ceramic protection tubes due to their lower thermal conductivity.

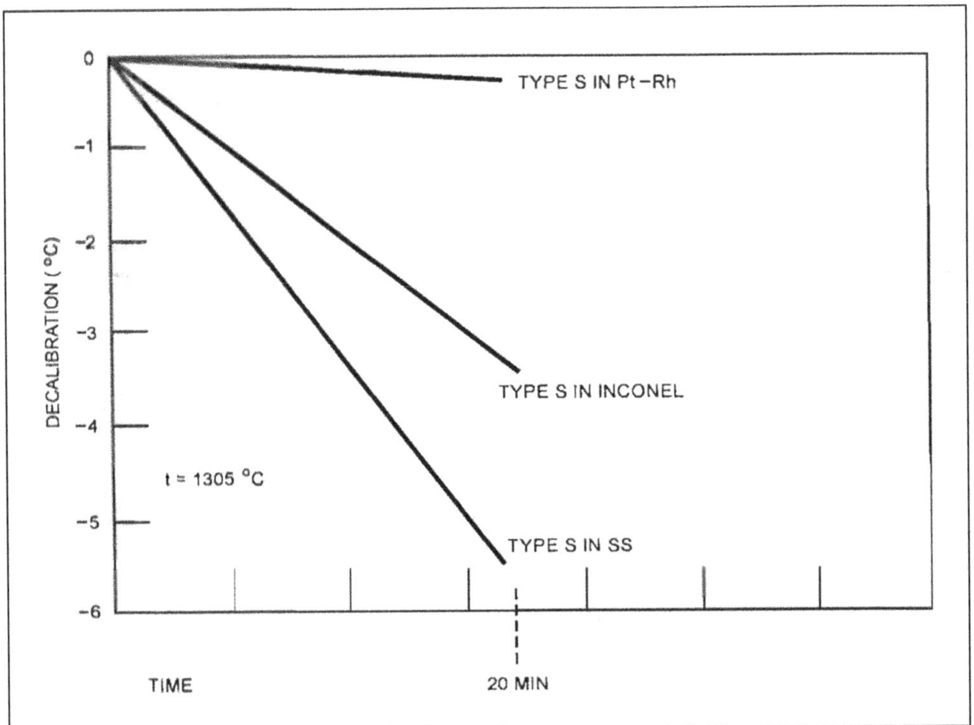

Figure 2-8. Incompatible metal sheaths are a source of contaminants that increase decalibration drift from composition changes in the thermoelements [14]

For resistance temperature detectors, decalibration errors will develop but usually to a lesser degree. The principal causes are strain, moisture, and grain growth at temperature extreme [24].

Temperature cycling at cryogenic temperatures or above 900°F alters the length and diameter of the RTD sensor coil and work hardens the platinum. Repeated cycling at these temperatures can cause resistance changes equivalent to several degrees. The error can be reduced by a construction that facilitates relative motion between the coil and its support [24].

Humid air and the breathing of the RTD probe through the seals and spaces between the insulation and wires at the back end results in water vapor being trapped inside the sheath. Above 1000°F, water molecules begin to dissociate into oxygen and hydrogen atoms, which are absorbed and further catalyzed by platinum. The increase in resistance can cause errors of several degrees. The error can be prevented by hermetically sealed probes or dehydration by heating the probes for several hours prior to commissioning [24].

Grain growth in platinum occurs at temperatures above 1200°F. The grain boundary eventually spreads across the entire diameter of the sensor coil. It weakens the metallic structure and makes the sensor susceptible to failure from vibration and mechanical or thermal shock [24].

2-9. Insulation Error

At high temperatures, the insulation resistance can decrease enough for an electrical shunt to develop between the sensor and its sheath. The resistance of magnesium oxide decreases at an average rate of about one decade per 200 degrees between 50°F and 2500°F. For small diameter sheathed RTDs, the initial resistance at ambient temperatures could be as low as 10^7 ohm, and an error of 40° would occur at 1200°F from an insulation resistance of 10^4 ohm effectively in parallel with the sensor resistance [25]. The detected resistance continues to decrease with temperature and can create a dangerous situation for an exothermic reaction [24]. The resistance between the conductor and sheath (R_i) can be measured and substituted into Equation 2-20 to estimate the error from electrical shunts. For DIN platinum RTDs, a temperature coefficient (dR/dT) of 0.214 ohm per degree Fahrenheit (0.385 ohm per degree Centigrade) can be used. The problem increases with length from an increase in the equivalent number of shunts and humidity due to the hydroscopic tendencies of magnesium oxide. In one application, exceptionally long slender RTD probes for a fluidized bed reactor on the Gulf Coast exhibited errors of 100° after start-up.

$$E_9 = (R_s)^2/[(R_i + R_s) * (dR/dT)] \tag{2-20}$$

where:

E_9 = temperature error due to RTD insulation resistance (°F)

R_i = insulation resistance (ohm)

R_s = sensor resistance (ohm)

dR/dT = temperature coefficient of RTD (ohm/°F)

Virtual junctions for thermocouples develop from electrical shunting. For sheathed type K thermocouples with diameters of one millimeter above 1500° and all sizes of sheathed tungsten – rhenium thermocouples above 2550°F, electrical shunting has been observed for magnesia insulation. Errors of one to ten percent of the temperature reading occurred [25]. For error accumulation, the insulation error for operation above 2500° can be approximated as the tolerance specification.

For thermocouples above 2500°:

$$E_9 = E_6 \text{ (sensor tolerance)}$$

2-10. Leadwire Error

For RTDs, it is rather straightforward to estimate the temperature error from an uncompensated leadwire resistance. It is simply the leadwire resistance divided by the temperature coefficient, as shown in Equation 2-21, where again a temperature coefficient *(dR/dT)* of 0.214 ohm per degree Fahrenheit can be used for a DIN platinum RTD. For example, 500 feet of 20-gage leadwire has a resistance of 10 ohm and causes a temperature error of 47°F. The use of a three-wire circuit would compensate for this resistance and reduce the error to that associated with the difference in resistances between the two leadwires. Since the tolerance of most conductors is 10 percent, a temperature error of 4.7° could still exist [24]. A four-wire circuit would compensate for the leadwire mismatch in resistance and essentially eliminate the temperature error.

$$E_{10} = R_w/(dR/dT) \qquad (2\text{-}21)$$

where:

E_{10} = temperature error due to RTD leadwire resistance (°F)

R_w = resistance of lead wire (ohm)

dR/dT = temperature coefficient of RTD (ohm/°F)

The extension leadwire for thermocouples has an ISA tolerance specification that is about the same as for the sensor, except for extension wires for platinum-type thermocouples where alternate alloys have to be used to keep the cost reasonable. The error for type SX is +12 and for type BX is +59°F [14]. Also, there are no special grade options for the extension wire. Table 2-11 shows the tolerances and associated temperature ranges for various types of thermocouple extension wires.

Table 2-11. The ISA extension wire tolerances are large for the alloys used for the platinum type thermocouples [14]

TC Type	Wire Type	Alloys Typical	Range, °F	E10 °F, Standard	E10 °F, Special
E	EX	Ni-Cr/Constantan	0 - 204	3.1	—
J	JX	Fe/Constantan	0 - 204	4.0	2.0
K	KX	Ni-Cr/Ni Alloy	0 - 204	4.0	2.0
T	TX	Cu/Constantan	–53 - 93	1.4	0.7
R,S	SX	Cu/Cu-Ni Alloy B	0 - 204	12.1	—
B	BX	Cu-Mn Alloy/Cu	0 - 121	59.4	—

Each termination of the extension wire creates a virtual junction and introduces another error into the system. The junction from the extension wire connection in the thermocouple head (E_{11}) and in the instrument (E_{12}) create additional errors that can be approximated as each are equal to the tolerance specification [23]. Excessive thermal insulation around the thermocouple head can cause its temperature to be outside the range shown for extension wires and can create a larger junction error.

For thermocouple terminations in the head and instrument:

$$E_{11} = E_{12} = E_{10} \text{ (wire tolerance)}$$

An error is also created from lead wire resistance for thermocouples that depends upon the input impedance of the electronics. Figure 2-9 plots the error in percent of reading versus leadwire resistance for three different circuit input impedances. For example, the resistance of 1000 feet of 20 AWG type KX extension wire is 590 ohm and creates an error of 1.18 percent for a 50,000-ohm input impedance [26].

2-11. Error Accumulation

The total measurement error can be approximated per Equation 2-22 as the summation of the systematic errors with the quadrature (root mean square) combination of tolerance limits and random errors [27]. The systematic errors are the conduction (E_1), dynamic error (E_2), radiation error (E_3), and velocity error (E_4) that are associated with the probe design and process installation and constitutes the difference between true process temperature and sensor temperature. For a well-designed pipe elbow installation without abrupt temperature changes or radiation or velocity error considerations, the systematic error is probably less than one percent of reading and is mostly due to fluid temperature profile. For less than ideal installation details and operating conditions, this error rapidly escalates to ten percent or more.

The tolerance limits and random errors are the errors E5 through E12 that are associated with the instrument system (sensor, wiring, and instrument). The accumulated instrument system error for a thermocouple is between three and five times the tolerance limit for the sensor. For example, the total instrument system error was 28.8° for a type K thermocouple with a calibration tolerance of 6.8° at 1800°F [14]. The accumulated instrument system error for a standard size industrial PRTD below 1200°F with a four-wire circuit can be approximated as between one and a half and two times the tolerance limit.

$$E_m = E_p + E_i \quad (2\text{-}22a)$$

$$E_p = E_1 + E_2 + E_3 + E_4 \quad (2\text{-}22b)$$

$$E_i = (E_5^2 + E_6^2 + E_7^2 + E_8^2 + E_9^2 + E_{10}^2 + E_{11}^2 + E_{12}^2)^{0.5} \quad (2\text{-}22c)$$

For industrial 4-wire PRTDs in good condition:

$$1.5 * E_6 < E_i < 2.0 * E_6$$

For industrial TCs in good condition:

$$3.0 * E_6 < E_i < 5.0 * E_6$$

where:

E_p = systematic errors from process installation (°F)

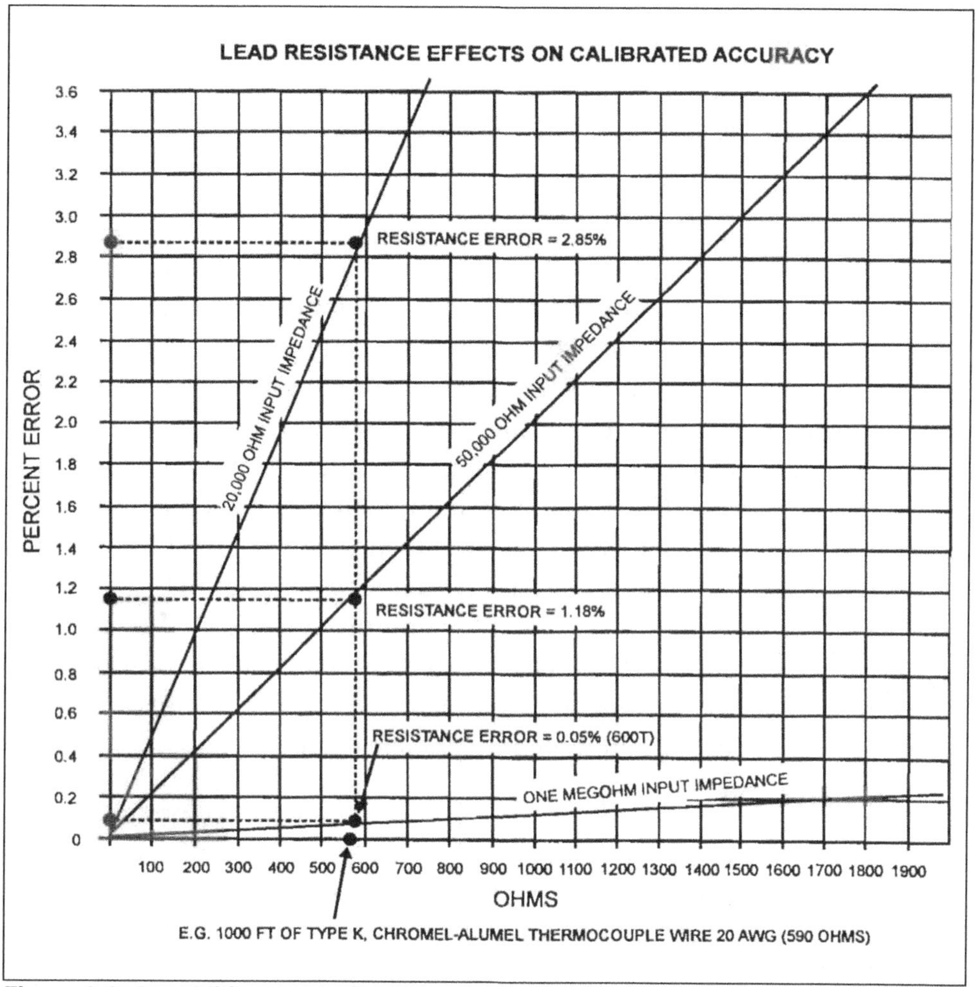

Figure 2-9. An additional error for thermocouple extension wire depends upon the circuit input impedance [26]

E_i = tolerances and random error from instruments (°F)

E_m = accumulated measurement error (°F)

E_1 = heat conduction error (°F)

E_2 = dynamic error (°F)

E_3 = radiation error (°F)

E_4 = velocity error (°F)

E_5 = electronic error (°F)

E_6 = sensor error or tolerance limit (°F)

E_7 = nonlinearity error (°F)

E_8 = decalibration error (°F)

E_9 = insulation error (°F)

E_{10} = leadwire error (°F)

E_{11} = head termination error (°F)

E_{12} = instrument termination error (°F)

It is important to note that the sensor tolerance limit (E_6) is a goal. Furthermore, compensation for this tolerance limit by calibration doesn't eliminate sensor error. In real life, mistakes and quality control problems during the manufacture of the sensors can result in larger than expected changes in calibration or even failure of the sensor in service. Tests of 102 PRTDs and 66 TCs, donated by manufacturers, cycled between 390 and 570°F, revealed some interesting statistics. Most of the PRTDs changed calibration by 0.4° and all of the TCs changed calibration by at least 2.5°F. Furthermore, four of the PRTDs went through calibration changes of 5°F to 36°F and fourteen failed. The greatest change in calibration for the TCs was 25°F. These results indicate the need for redundant sensors and sensor validation by automated diagnostics [28].

In a separate series of tests of 1/8-inch sheathed magnesium oxide-insulated thermocouples in the severe chemical and thermal environment of underground fossil fuel retorts, the failures of several of the thermocouples were observed with special online validation diagnostics. None went to the open circuit condition normally thought of as the failure mode. Thus, simple burnout detection for thermocouples cannot be relied upon for failure protection. Also, some thermocouple failures showed no noticeable change in temperature reading or noise. It was concluded that the best diagnostic was a continual high resolution measurement of changes in loop resistance, and that it was unwise to depend on thermocouple measurements in critical applications without diagnostics [29].

2-12. New Sensors

Johnson Noise Thermometers (JNT) are being developed for a nuclear reactor in a space station that provide reliable unmaintained operation with a measurement uncertainty of less than one percent for seven to ten years at temperatures greater than 1800°F. The sensor accuracy is not degraded by changes in insulation or sensor resistance, provided the insulation resistance is greater than 1000 ohm and the sensor resistance is between 10 and 1000 ohm. The sensor incorporates a four-wire RTD, as shown in Figure 2-10, for measurement comparison and the loop current step response (LCSR) method to determine the speed of response of the sensor. The JNT circuit computes the temperature from high frequency, low level, stochastic voltage signals. It is susceptible to nonthermal noise sources such as electromagnetic interference (EMI) [30].

While the JNT is still in the development phase, the optical fiber thermometer (OFI') is in the start of its exploitation phase. The OFT has proven to be critical to improved temperature control for difficult applications in steel mills, fiberglass factories, and semiconductor processing. It has proven to survive extreme mechanical stresses in auto engines, turbines, and rocket exhausts. The OFT has overcome early detractors such as cost per measurement point and sensor length, durability, coating, and installation. Figure 2-11 shows the different sensor options. The light pipe sensor can be used to 7200° if the sapphire sensor temperature does not exceed 3600°F [31].

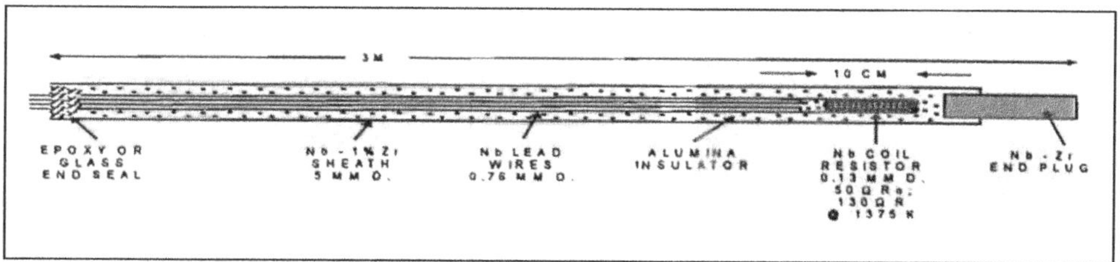

Figure 2-10. The Johnson noise thermometer uses a four-wire RTD for signal validation and dynamic response verification [30]

The OFT eliminates the normal problems associated with optical pyrometers such as variable emissivity and transmission factors, because the target is the tip of the fiber probe so that the geometry, surface conditions, and transmission medium are fixed.

The OFT has a resolution of 0.02°F, a repeatability of 0.01 percent, and a drift of less than 0.2°F per year. OFTs, combined with a PID controller with a high resolution, sixteen-bit analog/digital convertor, have proven to control furnace temperatures within 0.05°F. This is much tighter control than can be achieved with thermocouple systems, mostly because of the adverse effect of a thermocouple's thermal noise, drift, and susceptibility to EMI [31].

The thermal noise of a precious metal thermocouple at 1800° has been recorded, as shown in Figure 2-12, to have a 25-hertz bandwidth and peak-to-peak amplitude of one degree Fahrenheit. Thermocouples are also susceptible to electromagnetic interference (EMI). This is generally not the case for an OFT.

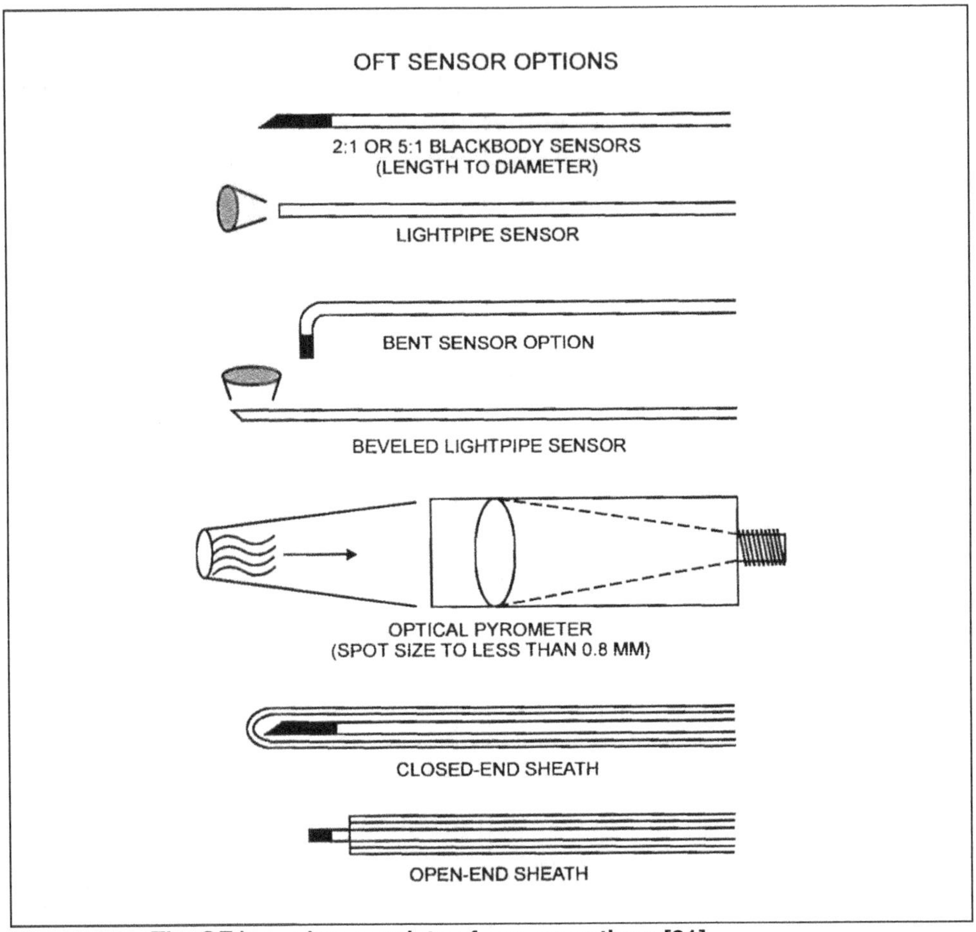

Figure 2-11. The OF has a large variety of sensor options [31]

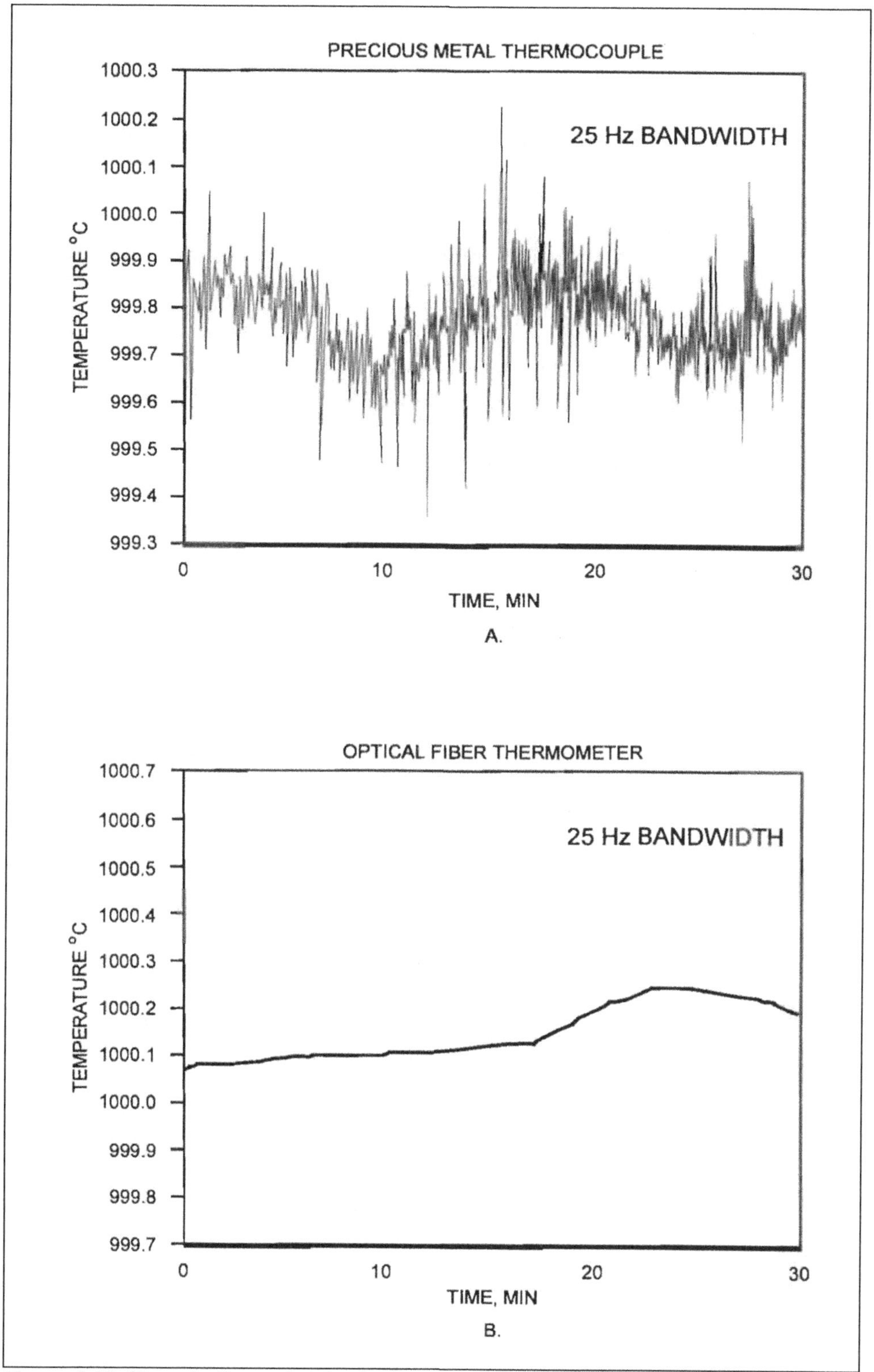

Figure 2-12. There is thermal noise in the signal from precious metal thermocouples due to their fundamental physical principles that isn't found with an OFT [31]

Exercises

2-1. What are the two major sources of thermal resistance that contribute to the heat conduction error caused by a thermowell? In order to reduce the heat conduction error, which of these thermal resistances should be larger?

2-2. Using Equation 2-2 calculate the thermal resistance between the interior tip of an 18-inch 316 stainless steel thermowell and its process connection. The thermowell's wall thickness is .25 inches with an inner diameter of .25 inches.

2-3. A 9-inch Hastelloy B thermowell is to be used in a polymer service with a convective heat transfer coefficient of 10 Btu/hr * ft^2 * °F. Using spreadsheet software, Equation 2-3 and Appendix H determine the maximum diameter of the thermowell for .4% conduction error.

2-4. What is the best location and type of thermowell installation and why is it the best?

2-5. What actions can be taken to prevent heat radiation errors?

2-6. Using Equation 2-9 and Appendices E and H find the temperature error due to radiation for a 316 stainless steel thermowell placed inside a superheated steam pipe. The outside surface temperature of the probe is 450°F and the interior pipe temperature is 1000°F. The probe is 1 inch diameter and 10 inches long.

2-7. After applying a curve fitting program to temperature data of a reactor, an equation was developed for the way the reactor heats up under a constant steam load. The equation is:

$T = .03t^2 + 10t + 22$

T= temperature °F

t = time in minutes

Using Equation 2-10 determine the dynamic temperature error at 12 minutes if the measurement time constant is 45 seconds.

2-8. Using Equation 2-14 calculate the total temperature error developed when hot air at 150°F is blown by at a thermocouple perpendicular to the flow at 600 ft./sec. The speed of sound of 150°F air at standard atmospheric pressure is 1211 ft./sec.

2-9. Which type of temperature sensors have the best repeatability: RTDs, thermocouples or thermistors? Which have the best sensitivity? Which have the highest temperature range?

2-10. A reactor temperature loop in a DCS system is tuned with the following settings: gain = 5, reset = .1 minutes/repeat, rate = .8 minutes. The DCS has a built in filter of 1/8 of the derivative setting. Using Equation 2-16, determine which controller setting would have the biggest effect on controller output if it were doubled?

2-11. What is the primary cause of decalibration in thermocouples and in RTDs?

2-12. Name the twelve types of temperature error. Which of these are a function of the way the sensor is installed in the process and which are a function for the way the sensor was manufactured?

References

1. Carter, D. E, "Problems and Solutions in Polymer Melt Temperature Measurement," (Monsanto Paper).
2. Crawford, C. L, "Thermowell Heat Conduction Error Versus Length." *Proceedings of Texas A&M Instrumentation Symposium*, 1982.
3. Richmond, D. W., "Selecting Thermowells for Accuracy and Endurance," *InTech*, February, 1980.
4. Behrmann, W. C., "Thermocouple Error Due to Sheath Conduction," *InTech*, August, 1990.
5. Burke, G. P., "Optimizing Severe Service Thermowell Design," *Power Engineering*, August, 1981.
6. Masek, J. A., "Guide to Thermowells," *Temperature Developments*, Omega Engineering, Inc., Stamford, CT.
7. Stroik, E. R., "RTD's are Sturdier Than You Think," *Instruments & Control Systems*, June, 1980, pg. 28.
8. Bolles, W. L, "Measurement of Gas Temperatures by Means of Thermocouples," *Petroleum Refiner*, February, 1948.
9. Mullikin, H. F., and Osborn, W. J., "Accuracy of the High Velocity Thermocouple," *Temperature—Its Measurement and Control in Science and Industry*, Reinhold Publishing Co., 1941, P 80S.

10. Buckley, P. S., "'Dynamics of Temperature Measurements," Process Annual Symposium, Instrumentation Process Industry, Wilmington, Vol. 34, 1979, pp. 21- 29.

11. McMillan, G. K., *Tuning and Control Loop Performance*, third edition, ISA, 1994.

12. Kardos, P. W., "Response of Temperature Measuring Elements," *Chemical Engineer*, August 29, 1977.

13. "Platinum Resistance Temperature Sensors," Rosemount Bulletin 9612 Rev. B, 1962, pg. 5.

14. Anderson, R. L.; Adams, R. K.; and Duggins, B. C., "Limitations of Thermocouples in Temperature Measurements," ISA, 25th International Symposium, Anaheim, May, 1979.

15. Bediones, D., and Wang, T. P., "Criteria for the Selection of Thermocouples Versus RTDs in Industrial Applications," ISA, International Conference, Anaheim, October, 1991.

16. Loeffler, R. F., "Thermocouples, Resistance Temperature Detectors, Thermistors—Which?," *Instruments & Control Systems*, May 1973, pg. 93.

17. Arndt, D. M., "An Introduction to Radiation Pyrometry," Ircon Bulletin, pp. 224-3.

18. Tenny, A. S., "Applying Radiation Thermometers to Process and Lab Measurements," *InTech*, August. 1988, pg. 51.

19. "Platinum Resistance Surface Temperature Sensors," Rosemount Bulletin 1011, 1973, pg. 3.

20. "Guide to Thermocouple Temperature Measurement," Fischer & Porter Bulletin 12-1, May, 1973, pg. 3.

21. "Practical Temperature Measurements," Hewlett Packard, Application Note 290, August, 1980, pp. 7 - 11.

22. Wang, T. P., and Bediones, D., "Improvement of High Temperature Stability of Thermocouple in Air via Metal Sheathing and Ceramic Fibre Coating," ISA, International Conference and Exhibit, Houston, October, 1988, pp. 931-942.

23. Kennedy, R. H., "Selecting Temperature Sensors," *Chemical Engineering*, August 8, 1983, pp. 54-71.

24. Trietley, H. L., "Avoiding Error Sources in Platinum Resistance Temperature Measurements," *InTech*, February, 1982, pp. 57-60.

25. Kerlin, T. W., and Shepard, R. L., *Industrial Temperature Measurement*, ISA, Instructional Resource Package, 1982, pp. 125-129, 159-161.

26. Rosenberg, R. J., "How Accurate are Your Process Temperature Measurements?," ISA International Symposium, 1975, paper 827, pg. 4.

27. Howard, J. Lawrence, "Error Accumulation in Thermocouple Thermometry," Specialized Techniques in Thermocouple Thermometry, Paper 186.

28. Kerlin, T. W., and Katz, E. M., "Temperature Measurement in the 1990s," *InTech*, August, 1990, pg. 41.

29. Reed, R. P., "Validation Diagnostics for Defective Thermocouple Circuits," 6th Symposium on Temperature sponsored by the American Institute of Physics, National Bureau of Standards, and ISA, Washington, DC, March 14-18, 1982.

30. Shepard, T. V., et al., "Development of a Long-Life, High-Reliability, Remotely Operated Johnson Noise Thermometer," ISA International Symposium, October, 1991, Anaheim, paper 91-0306.

31. Tinsley, F. G., and Adams, B., "Evolution in the Application of Optical Fiber Thermometry (OFT)," ISA International Symposium, October, 1991, Anaheim, Paper 91-0307.

3
Basic Feedback Control

Learning Objectives

A. Be able to select the best PID structure and form for an application.

B. Know how to tune PID controllers using a unified approach.

C. Recognize how diverse tuning methods can be reduced to a common form for tight control.

D. Discover how to reduce the process test time by an order of magnitude.

E. Recognize how to prevent fast oscillations caused by aggressive tuning and slow cycling and offsets caused by sluggish tuning.

F. Understand the implications of an integrating response on PID tuning.

G. See how an adaptive control can provide process knowledge in addition to automatically identifying tuning settings.

H. Find out how to significantly speed up the set point response to reduce batch time.

3-1. Introduction

Adding feedback measurement is essential for important process outputs because of the uncertainties and variability both in the process inputs and within the process. Using measurement in a feedback control loop offers automatic compensation. Since the proportional-integral-derivative (PID) controller is the predominant controller used in industry for basic feedback control, this chapter focuses on how to set up, tune, and optimize the PID.

Chapter 4 discusses how a control loop's ultimate performance depends on a process model but that the actual performance is determined by the PID controller's tuning settings. This chapter details the procedure for computing the PID tuning settings from the parameters of a process model. Though a dynamic model is implied in the tuning, as long as the PID controller is stable, it corrects for unknowns, loads, and disturbances.

In general, the user backs off from the "hottest" tuning settings for the "tightest" control in order to reduce the potential for oscillations in the process variable or manipulated variables. A trade-off must always be made between performance and robustness. High controller gains transfer more variability from the process variable to the manipulated variable. Fortunately, column, reactor, and vessel loops have fewer interactions than do many other unit operations, so variability in the manipulated variable is less disruptive. However, fluctuations in the controller output as a result of process or measurement noise that exceeds the final element's resolution can disturb the loop. Aggressive control makes the loop more vulnerable to oscillations from the inevitable changes in the process dynamics. Also, operators tend to dislike the large kicks in the controller output that can occur from set point changes to controllers with a high gain even though these kicks may in fact be beneficial in terms of a creating a faster response.

This chapter shows how process dynamics permit controller gains for the primary loops that are higher than those users have used on loops in other unit operations. In practice the controller gains actually used are far below the optimum for tight control because of the user's comfort zone combined with concerns about robustness and amplification of noise.

The oscillations caused by hot controller tuning and noise amplification are relatively fast and better understood than are the slower oscillations from less obvious causes. This chapter details how sluggish tuning of secondary controllers causes cycling in a cascade control system, how low controller gains cause reset cycling, and how sluggish tuning settings aggravate the observed limit cycling from the final element's resolution limit.

PID controllers are commonly used in temperature control. Before tuning tests are done and tuning decisions made, users need to select the proper structure and form of the PID algorithm. The effect that any set of tuning settings will have on loop performance depends on the PID's structure and form. A modern distributed control system (DCS) offers two or more forms and eight or more structures. It's important for users to understand the functional capability of each choice in order to standardize on the best algorithm for the application. When migrating projects or relocating settings from laboratories or pilot plants to production units, it is important to know the effects of different controller algorithms and the tuning parameter units. This knowledge enables users to reuse what they learned from tuning previous systems or similar applications in current systems.

Loop tuning and analysis tools also need to take into account tuning setting units, PID structure and form, and the limitations on the testing of production systems. The test times and step sizes are often beyond what is acceptable to plant operations. This chapter discusses how test time can be reduced by an order of magnitude for the primary loops, considerations for using the tuning settings from bench-top units and pilot plants, how to evaluate batch set point responses, and tuning methods that can be extended to provide optimal estimates of the tuning settings for industrial production. Ultimately, the user needs to understand the functional contribution of each mode in order to verify, analyze, and improve controller tuning and performance.

Section 3-2 describes the relative importance of various PID forms and structures; issues to consider when moving tuning settings between bench tops, pilot plants, and production units; and the important advantages offered by external feedback. Section 3-3 develops a unified approach for tuning controllers, discusses how tuning methods can be reduced to a common form, and details the impact of process dynamics on primary controller tuning.

Section 3-4 provides an overview of an adaptive controller that identifies the process model from normal set point or output changes and eliminates the need for manual or programmed test sequences. The adaptive controller computes settings based on the users' preference of robustness versus performance and users' concerns over the amplification of noise. The tuning settings identified from past batches can be used to schedule (predict) settings as a function of controller output or any process or computed variable such as batch time. Limits can be placed as necessary on the range of settings. Whether these settings are used or not, the models identified provide process knowledge and diagnostics. Changes in the process model can alert operations staff to changes in column, reactor, and vessel behavior.

Section 3-5 compares conventional techniques for pre-positioning controller outputs and tuning controllers against a new innovative approach that considerably increases the speed of the primary loop's set point response through a simple calculation of the controller output's optimal switching time and final resting value.

3-2. PID Modes, Structure, and Form

Basic feedback control is performed by a controller that has proportional, integral, and derivative modes. Except for temperature loops, the derivative mode is usually turned off by setting the rate time (derivative time) setting to zero. Users make a distinction here and call a controller with no derivative action a PI controller, but the open literature often does not do this. In many papers, the performance of PI controllers (labeled as PID controllers) is frequently compared to advanced control algorithms. This section will discuss how heavily performance comparisons depend on PID structure. Equations 4-1 and 4-2 in Chapter 4 showed how strongly performance depends on tuning settings. The authors of the technical literature choose structures and methods to prove the value of their new tuning methods or algorithms. In cases where derivative control is useful and noise and interaction are negligible, an aggressively tuned PID controller offers the best rejection of unmeasured disturbances at the input to a process [1] [2]. Often overlooked are the special techniques that can be readily added to the PID controller, such as batch preload and dead-time compensation via external reset (mentioned in this section) and the optimal switching of PID output to its final resting value (detailed in Section 3-5).

Proportional Mode Structure and Settings

The discrete contribution that the proportional mode makes to the controller output for the "standard" form of the PID algorithm is shown in Equation 3-1. The set point is multiplied by a β factor that ranges between 0 and 1 and is used to provide a proportional kick to speed up the response to a set point change. The kick is a step change in output. For slow loops, the step drives the output beyond its final resting value. Without this kick, the PID controller output relies on the integral action, which provides a slow approach to the set point by way of a slow reset time that is set to match the slow process response. The benefit the kick provides is greatest for loops in which the process time constant or integrating process gain is much slower than the process dead time. Thus, the kick provides minimal benefit to a flow loop since the process time constant is so fast. However, if the flow loop has a cascade set point or remote cascade set point, the kick can be used to provide a more immediate response to the demands of a primary controller or a batch sequence, as discussed in Sections 3-3 and 3-5, for cascade and batch control, respectively. However, whenever the operator changes a set point, this kicks the output. For a primary loop or a single loop, a set point rate limit or filter can be added to smooth out the kick from an operator set point or the β factor can be set to 0. It's not advisable to use set point rate limits and filters for secondary loops because they degrade the primary loop's performance. They have the same effect on the controller output as a velocity limit or filter. Although the β factor in a secondary loop degrades a primary loop's ability to reject disturbances, it has no deteriorating effect for disturbances that originate within a secondary or single loop.

A structure of "PI action on error," which sets the β factor = 1.0, provides the fastest set point response. This is important for secondary loops and the batch sequences.

In the equations for the PID controller used in this book, we use the term *controlled variable* (CV) in place of *process variable* (PV) both to denote that the units are percent of scale range instead of engineering units of the process variable and to make the nomenclature for the PID and model predictive controller (MPC) more consistent. It is important to remember that the configuration, displays, trend charts, and documentation of most modern control systems use process variable (PV), feedforward variables, and controller outputs in engineering units. When computing the process gain, users must take this into account because in a distributed control system (DCS) the PID algorithm is based on feedback and feedforward inputs and an output in percent of the respective scales. Special-purpose and user-constructed PIDs, as well as a few programmable logic controllers (PLC), use engineering units in the algorithm. This severely reduces the portability of tuning settings and the users' understanding of the effects of the process. Section 3-3 on PID tuning methods shows that the change in engineering units cancels itself out for PID algorithms that use percent for the controller inputs and outputs. The specification of engineering units for controller output in modern DCS and fieldbus systems does not mean these engineering units are used in the PID algorithm. The conversion of controller output from percent to engineering units is done after the PID algorithm.

The use of percent for inputs and outputs in the PID control algorithm increases the portability and comprehensibility of controller gain settings.

$$P_n = K_c * (\beta * SP_n - CV_n) \qquad (3\text{-}1)$$

Though most digital controllers use controller gain, proportional band (PB) was once a prevalent tuning setting for the proportional mode in analog controllers. Proportional band was devised to be the percent change in the controlled variable that is needed to cause a 100 percent change in the controller output [2] [3]. Some digital controllers still use it. It is critical that the user know whether the proportional mode tuning setting is a gain or proportional band because this can have a huge effect as indicated by the inverse relationship shown in Equation 3-2.

It is critical to know whether the proportional mode tuning factor is a proportional band in percent (%) or a dimensionless controller gain.

$$PB = 100\% \,/\, K_c \qquad (3\text{-}2)$$

Integral Mode Structure and Settings

The discrete contribution of the integral mode to the controller output for the "standard" form of the PID algorithm is shown in Equation 3-3. The controller gain setting divided by the integral (reset) time setting is multiplied by the error between the set point and the controlled variable for the current execution (n). This result is multiplied by the integration step size (Δt), which is the module execution time, and added to the integral mode result for the last execution (n–1). The result is an integration of the error factored by the ratio of the controller gain to reset time setting. The reset time is the time required for the integral mode to repeat the contribution from the proportional mode. In the ISA standard algorithm, the reset time setting is in seconds per repeat and is simply called "reset." Often the units are stated as just seconds. Some manufacturers use minutes per repeat, or its inverse (repeats per minute), with no change in the use of the term reset. It is imperative that the user know the units of the reset tuning setting.

It is critical to know whether the integral mode tuning factor is a reset time (seconds or minutes per repeat) or its inverse (repeats per second or minute).

$$I_n = (K_c / T_i) * (SP_n - CV_n) * \Delta t + I_{n-1} \tag{3-3}$$

Derivative Mode Structure and Settings

The discrete contribution of the derivative mode to the controller output for the "standard" form of the PID algorithm is shown in Equation 3-4. The set point is multiplied by a γ factor that ranges between 0 and 1 and can be used like the β factor to provide a proportional kick to speed up the response to a set point change. In this case, the kick is a spike or bump, whereas the kick from the β factor is a step. The effect is short term, and the burden is still on the integral mode to change the output enough to accelerate the process variable. Although this jump in the controller output can help the signal get through the dead band or resolution limit of a control valve, the better solution is to reduce backlash and sticktion by using a more precise throttle valve and digital positioner. The γ factor is normally set to zero to prevent the derivative mode from overreacting to operator-entered set points. A set point filter or velocity limit can be used just as it was to reduce the kick from the β factor for single and primary loops. The γ factor, like the β factor, does not affect a loop's ability to reject a load.

The γ and β factor do not affect the load response of a control loop.

Since the derivative mode provides a contribution that is proportional to the slope of the change, step changes and noise are disruptive. Most controllers have an inherent filter whose time constant is a fraction α of the derivative (rate) time setting; this ensures that the spike in the output becomes a bump. A typical value for α is 1/8 to 1/10. Note in the Equation 3-4 that for α and γ factors of zero, the numerator can be simplified to just the change in controlled variable multiplied by the controller gain and rate time. The denominator can be simplified to just the execution time. For this case, it is easier to see that the derivative mode provides a contribution that is proportional to the rate of change of the controlled variable ($[CV_n - CV_{n-1}] / \Delta t$].

It is critical to know whether the time units of the derivative mode tuning factor (rate time) is seconds or minutes.

$$D_n = \frac{(K_c * T_d) * [\gamma * (SP_n - SP_{n-1}) - (CV_n - CV_{n-1})] + \alpha * T_d * D_{n-1}}{\alpha * T_d + \Delta t} \quad (3\text{-}4)$$

Summary of Structures

The β and γ factors are sometimes called *set point weighting factors* and are usually found under the category or term structure in the *controller configuration*. A controller in which both factors are adjustable is called a two-degree freedom controller. Other structures have the β and γ factors set equal to 0 or 1. The user can also omit a mode entirely to get P-only, I-only, ID, and PD control with various assigned factors. PI control is achieved by simply setting the derivative (rate) time to zero. In general, the user must not set the controller gain equal to zero in an attempt to get I-only or ID control. Nor should the user set the integral (reset) time to zero in an attempt to get P-only or PD control. Note that using P-only or PD control requires that additional choices be made about how to set the bias and its ramp time. Table 3-1 lists eight choices offered by one major DCS supplier.

Table 3-1. List of major pid structure choices

1. PID action on error (β = 1 and γ = 1)
2. PI action on error, D action on PV (β = 1 and γ = 0)
3. I action on error, PD action on PV (β = 0 and γ = 0)
4. PD action on error (β = 1 and γ = 1)
5. P action on error, D action on PV (β = 1 and γ = 0)
6. ID action on error (γ = 1)
7. I action on error, D action on PV (γ = 0)
8. Two degrees of freedom controller (β and γ adjustable 0 to 1)

Normally a structure of "D action on PV" (γ= 0) is used to prevent spikes or bumps from set point changes but "D action on error" (γ> 0) can be a temporary fix to minimize the effect of valve backlash and sticktion for set point changes.

Algorithms

In the "standard" form of the PID used in most DCS applications, the contributions of the three modes are added to the initial value of the controller output (CO_i) set to provide a bumpless transfer to the execution of the PID algorithm. This initialization of the controller output occurs for transitions from modes in which the output is manually set, the output is tracking an external variable for cascade or override control, or the output is remotely set by batch sequence. Execution of the PID algorithm occurs for the AUTO (automatic), CAS (cascade), and RCAS (remote cascade) modes [2].

Equation 3-1 through 3-5 are a simplified representation of the "positional" algorithm that is predominantly used in industrial control systems [3]. An "incremental" or "velocity" algorithm used by special-purpose supervisory computers developed in the 1970s would compute the change in controller output from each mode for each execution, and add it to the full controller output from the last execution of the PID algorithm. The "incremental" algorithm inherently eliminates bumps, windup, and synchronization considerations. For supervisory computers, the last output was read back as the current set point so the recovery from a failure of execution or the communication link was smooth. However, these and other concerns are now addressed by the use of the back-calculate (BKCAL) feature and initialization calculations of Fieldbus functional blocks. Each control system supplier also has methods for preventing reset windup and coming off of the controller output limits for the "positional" algorithm. The "positional" offers the advantage of using external feedback for improved override control through more effective transitions between the selections of controller outputs. The "positional" algorithm also offers proportional-only (P-only) and proportional-plus-derivative (PD) control and a fixed bias. The "incremental" algorithm has inherent integral action and no fixed bias. Since the incremental algorithm is not recommended, this book will focus on the equations and function of the positional algorithm.

The "standard" form of a PID positional algorithm is the most common type used in a DCS and offers P-only and PD control with an adjustable bias.

$$CO_n = P_n + I_n + D_n + CO_i \tag{3-5}$$

where:

CO_i = controller output at transition to AUTO, CAS, or RCAS modes (%)

CO_n = controller output at execution n (%)

CV_n = controlled variable at execution n (%)

D_n = contribution from derivative mode for execution n (%)

I_n = contribution from integral mode for execution n (%)

K_c = controller gain (dimensionless)

P_n = contribution from proportional mode for execution n (%)

PB = controller proportional band (%)

SP_n = set point at execution n (%)

T_d = derivative (rate) time setting (seconds)

T_i = integral (reset) time setting (seconds)

α = rate time factor to set derivative filter time constant (1/8 to 1/10)

β = set point weight for proportional mode (0 to 1)

γ = set point weight for derivative mode (0 to 1)

Figure 3-1 shows the combined response of the PID controller modes to a step change in the set point (ΔSP). The proportional mode provides a step change in the controller output (ΔCO_1). If there is no further change in the SP or the CV does not respond, no additional change in the output occurs even though there is a persistent error (offset). The size of the offset is inversely proportional to the controller gain. Integral action ramps the output unless the error is zero. Since the error is rarely exactly zero, the reset is always driving the output. Even if there are no disturbances, reset action causes a continuous equal amplitude oscillation (limit cycle) as it moves the PID output and the process variable through the resolution limits of the final element and measurement, respectively. The probability of a controlled variable coming to rest exactly at set point is next to zero because of the quantization of the process input by the final element resolution and the process output by the measurement resolution. Thus, a PV

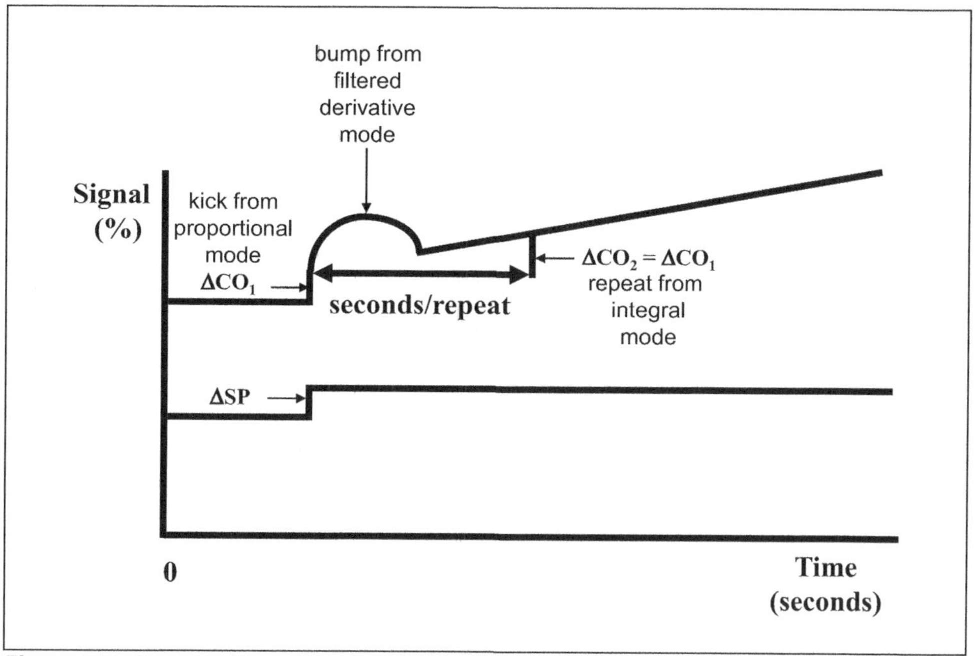

Figure 3-1. Contribution of each pid mode for a step change in the set point (β=1 and γ=1)

nearly always passes through the set point and never stays at a resting value for a controller that has integral action [2].

> ***The integral mode continually works to eliminate an offset and is always moving the PID output since the error is rarely exactly zero.***

The contribution made by the integral mode equals the contribution made by the proportional mode ($\Delta CO_1 = \Delta CO_2$) after the integral time. Hence, the integral (reset) time is the time (seconds) it takes to repeat the contribution of the proportional mode, which is the basis of the units "seconds per repeat," for a step change in the SP or CV [2] [3].

The contribution made by the derivative mode for the same step change is a bump rather than a spike because of the built-in derivative filter. If there is no further change in the SP and the CV does not respond, the contribution from the derivative mode goes to zero [2] [3]. Like the proportional mode, the derivative mode does not attempt to eliminate an offset. Consequently, proportional-plus-derivative (PD) controllers require that a bias be properly adjusted to minimize the offset, particularly for low controller gains.

Proportional-plus-derivative (PD) controllers with low controller gain settings require that attention be paid to the proper setting of the bias to ensure that the offset is acceptable.

Before the advent of the DCS, most industrial control systems used the "series" PID algorithm in which the derivative mode was computed first as the input to the proportional and derivative modes, as shown in Figure 3-2. The computation of the derivative mode in series with other modes was the most practical method for implementing derivative action in analog controllers and was known as the "real" algorithm. In analog controllers, the derivative mode with its built-in filter was actually a lead-lag in which the lead time was the derivative or rate time setting (T_d) and the lag time was the filter time ($\alpha * T_d$). In these analog controllers, the derivative action was on PV instead of error ($\gamma = 0$).

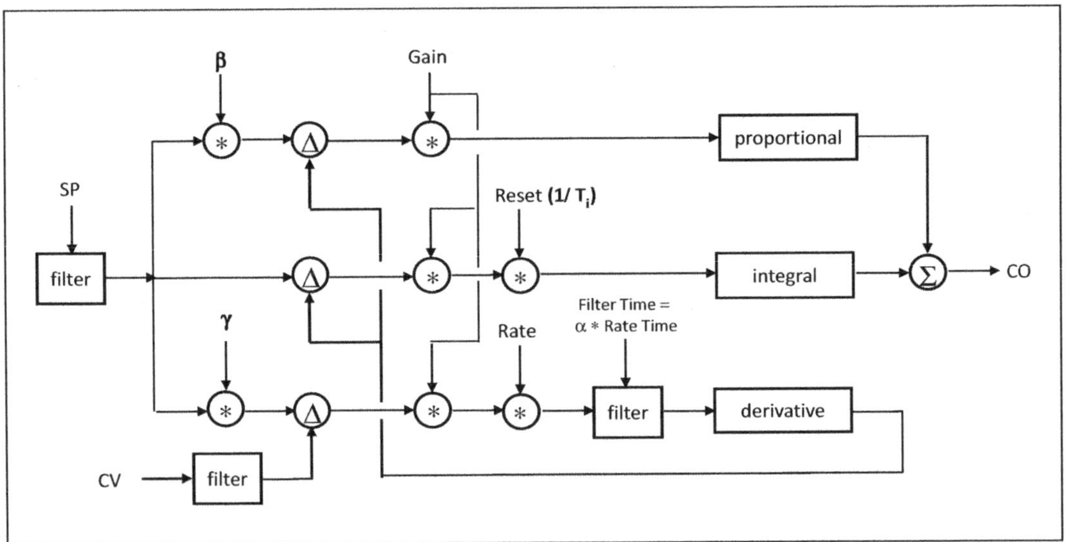

Figure 3-2. Block diagram of "series," "real," or "interacting" PID algorithm

The "series" algorithm was also known as the "interacting" algorithm because the derivative and integral time settings had an interacting effect on the contribution of all modes, as defined in Equations 3-6 through 3-9. From Equation 3-9, the interaction factor would ensure that the ratio of the derivative time to integral time never exceeded ¼. In theory, a ratio of ½ is optimal, but the prevalence of noise makes this goal impractical. A derivative time effectively greater than the integral time setting causes instability. If the user made a mistake and set the derivative time greater than the integral time, the "series" or "interacting" algorithm prevented the effective ratio from exceeding ¼ [3].

The "standard" algorithm computes the derivative mode in parallel with the other modes, as shown in Figure 3-3. It is the form of the PID algorithm adopted by ISA as its standard. The "standard" or ISA algorithm is also known as the "ideal" or "noninteracting" algorithm and is the default choice in most twenty-first-century control systems. The settings on the left side of Equations 3-6 to 3-8 are the settings for the standard algorithm. Note that when the derivative time is zero, the "standard" and "series" algorithm are the same, the interaction factor is 1.0, and no conversion of settings via these equations is needed [3]. Most of the literature on controller tuning assumes a "standard" structure. Hence, the derivative time is often listed as ¼ the integral time for PID control. In Section 3-3 on controller tuning, we discuss how derivative action is mostly used on temperature loops; it should be set equal to the next largest time constant (second lag). For thermal systems, the second lag is about 1/10 of the largest time constant.

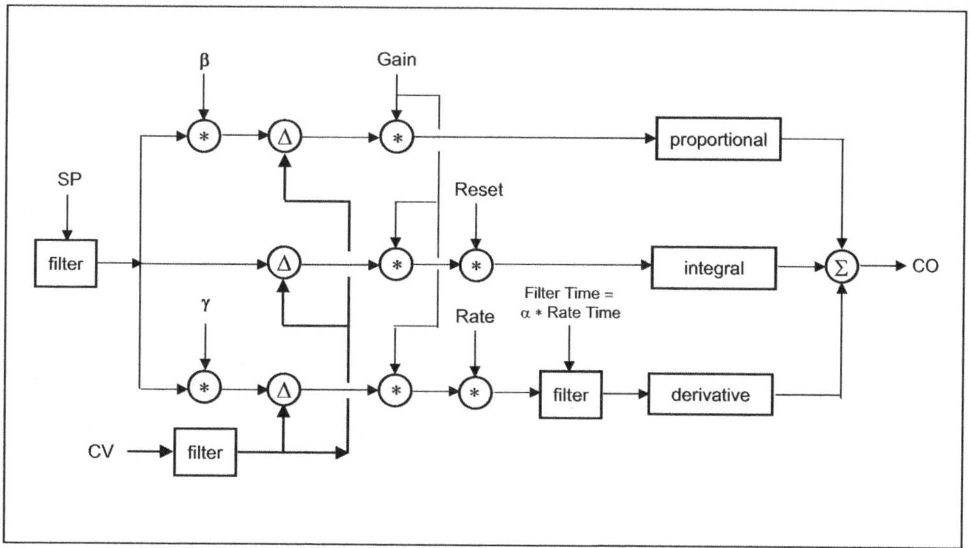

Figure 3-3. Block diagram of "standard," "ideal," or "noninteracting" PID algorithm

Much less common is the "parallel" algorithm, which also computes the modes in parallel. However, the controller gain setting only affects the contribution of the proportional mode. The controller gain is not in the input path to the derivative and integral modes. Only a few systems offer this as a choice. Although there is no performance incentive, some users may prefer that the tuning settings be completely independent. Other users are unappreciative of the drastically different "feel" and settings encountered when tuning a "parallel" controller. Also, the portability between the "parallel" and "standard" systems is much less than the portability between the "series" and "standard" systems. Unless the controller gain is close to 1.0, it is critical that Equations 3-10 through 3-12 be used to convert the settings between "parallel" into a "standard" controller when moving settings, regardless of whether derivative action is used.

As always, it is essential to take into account the different units of the tuning settings.

$$K_c = K_c' \;/\; I_f \tag{3-6}$$

$$T_i = T_i' \;/\; I_f \tag{3-7}$$

$$T_d = T_d' * I_f \tag{3-8}$$

$$I_f = T_i' \;/\; (T_i' + T_d') \tag{3-9}$$

$$K_c = K_c'' \tag{3-10}$$

$$T_i = T_i'' * K_c'' \tag{3-11}$$

$$T_d = T_d'' \;/\; K_c'' \tag{3-12}$$

where:

K_c = controller gain for "standard" algorithm (dimensionless)

K_c' = controller gain for "series" algorithm (dimensionless)

K_c'' = controller gain for "parallel" algorithm (dimensionless)

I_f = interaction factor (dimensionless number < 1.0)

T_i = integral (reset) time for "standard" algorithm (sec/repeat)

T_i' = integral (reset) time for "series" algorithm (sec/repeat)

T_i'' = integral (reset) time for "parallel" algorithm (sec/repeat)

T_d = derivative (rate) time for "standard" algorithm (sec)

T_d' = derivative (rate) time for "series" algorithm (sec)

T_d'' = derivative (rate) time for "parallel" algorithm (sec)

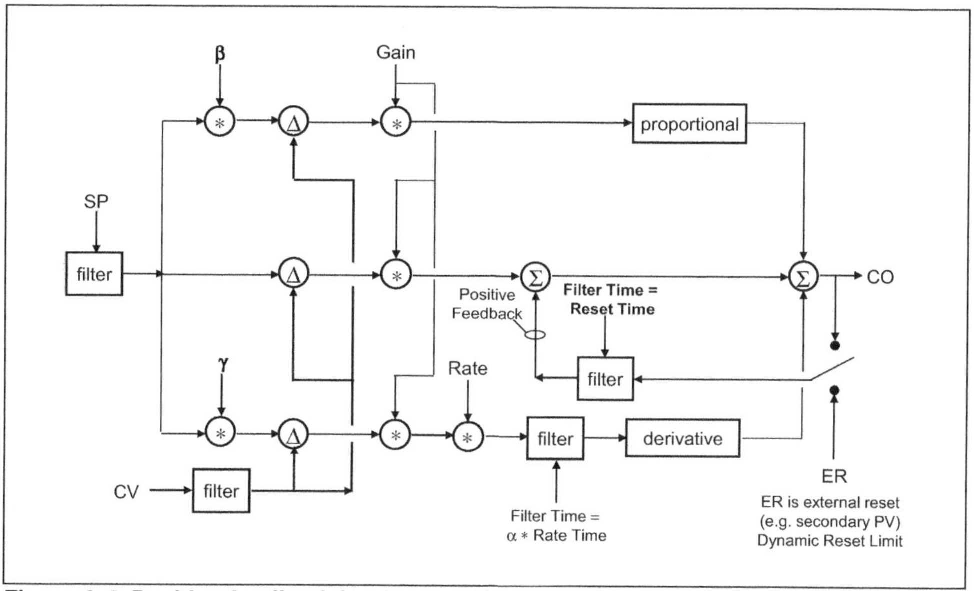

Figure 3-4. Positive feedback implementation of integral mode

Instead of an integration of the error for the integral mode calculation, some manufacturers have found it advantageous to use a filtered positive feedback, as shown in Figure 3-4 [4]. The input to the filter is either the controller output or an external feedback signal. The output of the filter is added to the net of the proportional and derivative modes in conformance to either the "series" or "standard" form. The filter time is the reset time setting. This positive feedback arrangement facilitates three important features [16]. First, the positive feedback makes it possible to use a "dynamic reset" option in which the external feedback is the PV of a secondary loop or control valve position. This option prevents the controller output from trying to go faster than the velocity-limiting effect of the reset time or the process time constant of the secondary loop, or faster than the slewing rate of a control valve. Second, the positive feedback arrangement inherently prevents a "walk-off" of controller outputs for override control. In the "walk-off," the controller outputs gradually move to an output limit from a continual back-and-forth selection of controller outputs when the integral mode uses an integration of error. The solution for these controllers is to add a filter to each external feedback signal with the filter time set equal to the respective controller's reset time. This effectively creates the same configuration offered by the positive feedback method when the controller is in the integral tracking mode. Third, a dead-time block can be added to the positive feedback path with its dead time set equal to the loop dead time. This provides a dead-time compensator as effective as a Lambda-tuned Smith Predictor but without the adjustments of process gain and lag.

The positive feedback type of integral mode prevents reset action from outrunning the speed of response of secondary loops and final elements, eliminates walk-off of override controller outputs, and facilitates simplified dead-time compensation.

3-3. PID Tuning

Focus

The wide spectrum of methods and results for tuning controllers can be bewildering. Most of the published literature analyzes methods for a particular range and type of process dynamics, disturbances at the process's output rather than its input, a set point response, fixed small dead times, and an accuracy of tuning settings not obtainable in industry. In general, the applications are high-speed servomechanism-type responses with measurement noise. Process disturbances enter into the process as changes in the charges, flows, or metabolic processes of the cell. These load upsets are process inputs that enter the process just as the manipulated variables do, as shown in the block diagram in Figure 2-2a of Chapter 2. Disturbances and the manipulated variables are process inputs that change the charge, component, and energy balances. The dead times and time constants associated with the primary loops' response are large and variable. Continual load upsets, the moving target of a batch profile, the nonlinearities enhanced by changing batch conditions, variable dynamics, resolution limits, and process noise from imperfect mixing place severe practical limits on the repeatability of tuning settings.

Control theory centers on high-speed servomechanism response with noise.

This section focuses on the Lambda method, showing how it can be adjusted to meet process goals and giving when desirable an equation similar to other tuning methods that are touted for loop performance. A translation of form also enables a much faster test time. The Lambda method has almost become a universal method because of its fundamental design.

The Lambda tuning method can be adjusted or transformed to meet any type of process objective and offers a unified approach to controller tuning.

Most of the articles to date on loop tuning are relevant to self-regulating processes with dead-time-to-time-constant ratios in the range 0.5 to 2.0, set point changes or step disturbances entering into the process output whose effect is similar to the set point changes for the most common type of controller (PI with $\beta=1$), and a presupposed accuracy of tuning settings of 10 percent or better. When integrating processes have been evaluated for tuning, it has typically been a level loop on a surge tank.

Most of the literature on tuning deals with a narrow range of dynamics, self-regulating processes, and a set point response for PI control action on error ($\beta=1$).

In the process industry, 99 percent of the self-regulating control loops have dead-time-to-time-constant ratios that range from 0.05 to 20.0. A true or "near integrating" type of response is prevalent in primary loops for concentration, pressure, and temperature control. The integrating process gain is usually extremely small and is equivalent to a very large process time constant. Process load upsets and disturbances occur at the process input instead of at the process output as commonly shown in the literature. Consequently, in industrial applications a process variable's rate of change from an upset or disturbance usually depends on the process time constant or integrating process gain. For columns, reactors, and vessels, the rate of change is very slow.

These same loops exhibit a variability of 50 percent or more in tuning settings from process tests as a result of noise, nonlinearity, resolution limits, and unmeasured disturbances [2]. A repeatability of 25 percent of tuning test results is considered exceptional. Realizing tuning setting precision in industrial applications eliminates a lot of the hype associated with methods and software. It also prevents false expectations, which is particularly important since engineers and scientists are accustomed to computing numbers to two or more digits [5].

Differentiating tuning methods and software based on the second digit of the tuning setting computation have little value in industry.

Temperature Loops

Temperature loops commonly have a process dead time and second time constant that are 0.01 to 0.1 of the largest time constant because of the interactive thermal lags [6]. A second-order model (two time constants) is useful for identifying the derivative time setting. However, in practice, it is difficult to accurately identify this second time constant so a first-order model (one time constant) is often used. This is often adequate for temperature loops since about half of the unidentified second time constant ends up as additional dead time and half ends up as an incremental increase in the largest time constant. The dead-time-to-time-constant ratio is about 0.01 to 0.1 whether a first- or second-order model is used.

The dead-time-to-time-constant ratio for most temperature loops on mixed volumes is about 0.01 to 0.1 whether a first-order or second-order model is used.

Because processes with large time constants and small dead times appear to ramp in the region of interest, the difference between a self-regulating and integrating process response is blurred and the distinction becomes a matter of convenience. If the process is visualized as integrating, the test does not have to wait until the process variable plateaus to a new final value. This is particularly advantageous for loops that have large time constants, since it takes about four time constants plus the dead time to reach steady state. If the user prefers to use tuning equations for self-regulating processes, the integrating process dynamics can be converted into self-regulating process dynamics, and vice versa.

The lack of a liquid discharge flow in a batch operation reduces the self-regulation of the temperature loop. However, for a reactor that has a well-designed coolant, the change in driving force across the heat transfer surface means that the temperature loop is not a pure integrator. For example, if the temperature change is large enough, the temperature difference between the process and coolant should be large enough to change the heat transfer to the coolant sufficiently for the temperature to reach a steady state. However, the temperature change may be much larger than the desired change in temperature and the time to reach a steady state too slow.

Reactor temperature loops have a time constant so large that the distinction between a self-regulating and integrating response is a matter of test time requirements and computational preferences.

Secondary Loops

In a cascade control system the output of the primary controller, such as column, reactor, and vessel temperature, is the set point of a secondary controller, which helps linearize the loop and compensate for disturbances that affect the final element's ability to do its job. Secondary flow and speed loops have a dead-time-to-time-constant ratio of between 1.0 and 4.0 unless a large signal filter has been added to measurement or there is significant velocity limiting in the drive or valve response. Most of the dead time in these loops comes from the module execution time and the resolution of the drive or valve response. Secondary coolant temperature loops have a much larger process dead time and time constant whose ratio and magnitudes significantly vary with the coolant system design.

Closed and Open Loop Responses

The terms *closed loop* and *open loop* are commonly used in control and are essential to understanding Lambda tuning. The term *closed loop* is used to denote that PID control action is active (PID is in AUTO, CAS, or RCAS modes). This means it is changing a manipulated variable in response to a process variable, effectively closing the signal path between the CV and CO of the PID in the block diagram of a control loop (see Figure 4-1 in Chapter 4). Conversely, the term *open loop* is used to denote that PID control action is suspended (PID in IMAN, MAN, or ROUT) and that the signal path is open between the CV and CO. The open loop response is the response of the process only. The closed loop response is the combined response of the process and PID and shows the effect of the control algorithm and tuning settings [2].

Figure 3-5 shows the open loop time constant (τ_0) that is created by making a step change in the controller output when the controller is in manual for a self-regulating process. Figure 3-6 shows the closed loop time constant known as Lambda (λ) that is created by making a step change in the controller set point when the controller is in automatic. In both cases, the time constant is the time it takes the controlled variable to reach 63 percent of its final value after the process variable starts to change (after the loop dead time). In tuning methods in which the time constant is estimated as

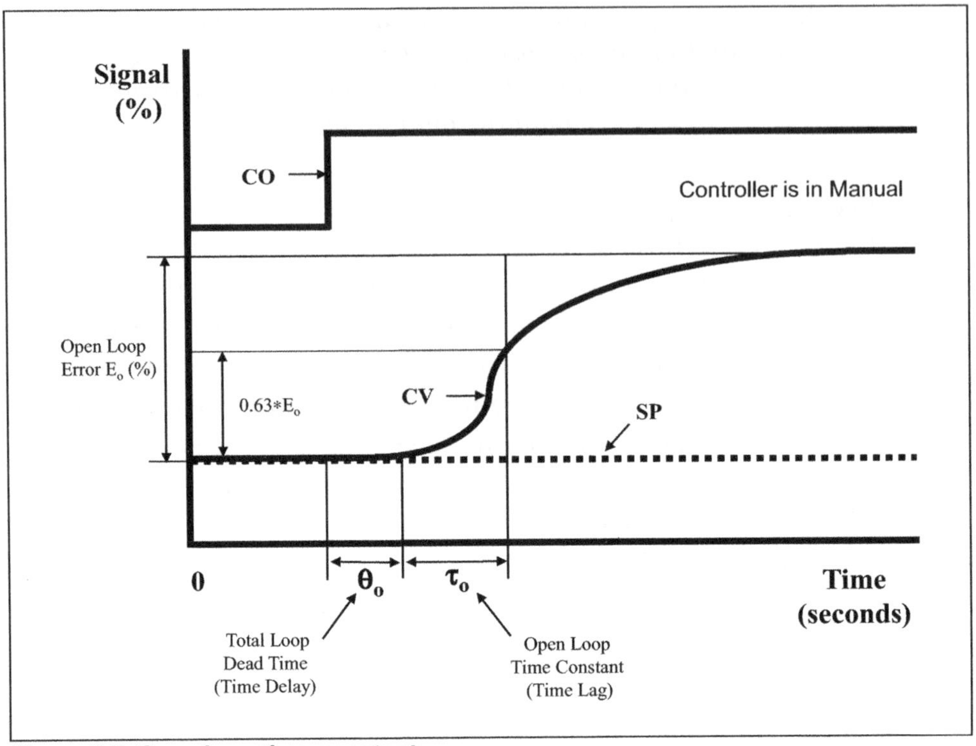

Figure 3-5. Open loop time constant

the time required to reach 98 percent of the final response divided by 4, the resulting time constant includes ¼ of dead time. In actual applications, there is no sharp transition from the negligible response caused by loop dead time and the exponential response associated with the primary time constant. The smoothing that occurs at the beginning of the exponential response is caused by a smaller second time constant (τ_2). Though this small time constant is difficult in practice to identify, it can be estimated for temperature and many concentration loops on well-mixed volumes as being 1/10 of the largest time constant.

The open loop time constant is the largest time constant in the loop. If it is in the process between the point of entry of the disturbances and the sensor, the open loop time constant slows down the excursion caused by a disturbance. This gives the controller a better chance of catching up to the disturbance. If the largest time constant is in the measurement, it provides an attenuated view of process variability and slows down the reaction to unmeasured disturbances. Coated electrodes can cause the sensor time constant to approach or even exceed the process time constant. Also, in the case of a liquid flow or speed loop, a transmitter damping setting or process variable filter time setting that is greater than two seconds generally

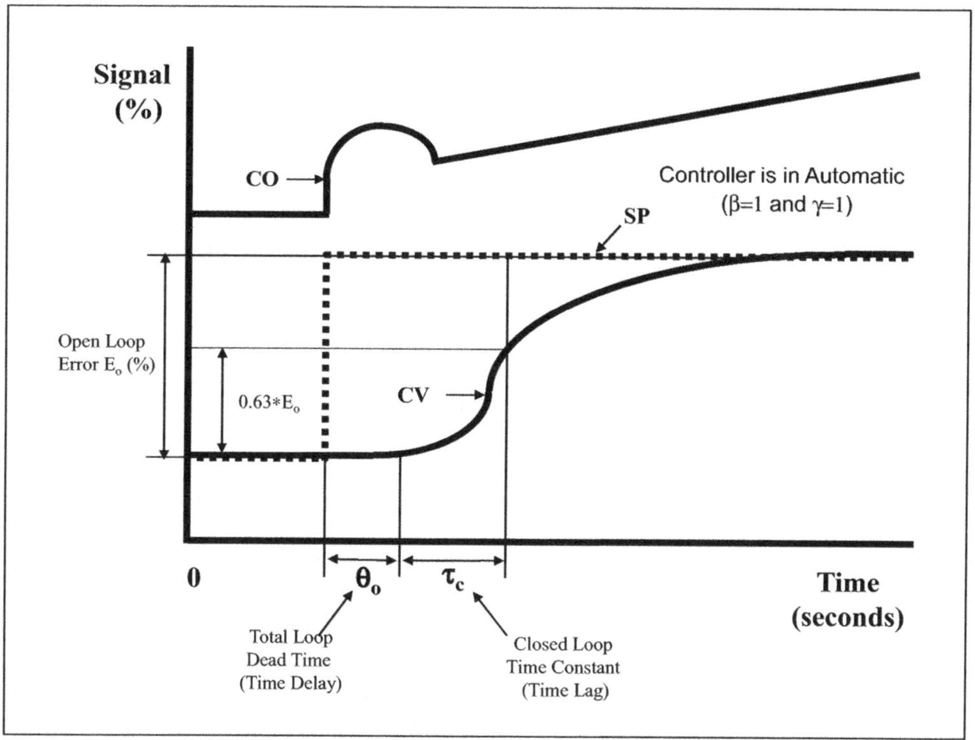

Figure 3-6. Closed loop time constant

means the largest time constant is in the measurement. This is because the liquid flow process and final element time constant is less than two seconds [2].

Lambda Tuning

The ratio of the closed-loop-to-open-loop time constant is the Lambda factor, as in Equation 3-13. For secondary flow and speed loops, a Lambda factor of 2 (λ_f =2) is a good starting point. This gives a closed loop time constant that is twice the open loop time constant. For primary loops, the Lambda factor is set less than one to provide tighter loop control.

$$\lambda = \lambda_f * \tau_1 \tag{3-13}$$

$$T_i = \tau_1 \tag{3-14}$$

$$K_c = \frac{T_i}{K_o * (\lambda + \theta_o)} \tag{3-15}$$

$$T_d = \tau_2 \tag{3-16}$$

where:

K_c = controller gain (dimensionless)

K_o = open loop gain (also known as process gain) (%/%)

λ = Lambda (closed loop time constant) (sec)

λ_f = Lambda factor (ratio of closed to open loop time constant) (dimensionless)

θ_o = total loop dead time (sec)

τ_1 = largest open loop time constant (also known as process time constant) (sec)

τ_2 = second largest open loop time constant (sec)

T_i = integral (reset) time setting (sec/repeat)

T_d = derivative (rate) time setting (sec)

For derivative (rate) action to be effective, the loop response should be smooth and have a significant second time constant [3]. This is generally the case for temperature loops on mixed volumes. Since the second time constant is about 1/10 of the largest time constant, the rate time setting for temperature loops is about 1/10 of the reset time setting.

Derivative action that has a rate time setting of about 1/10 of the reset time setting can provide tighter temperature control.

Contribution of Loop Components

The open loop gain is the product of the steady-state gains for each major component in the loop, that is, the final element (manipulated variable), process piping and equipment (process variable), and sensor and transmitter (controlled variable), as shown in Equation 3-17. The result must be dimensionless (%/%), and if it is not, a component is missing or the engineering units are not consistent. The manipulated variable gain is generally more linear for variable speed drives than for control valves. The process gain is usually nonlinear for temperature and concentration loops. The process gain is 1.0 for flow. The measurement gain is linear and provides a conversion from the engineering units of the process variable into percent of the measurement span (K_{cv} =100%/EUspan) [2] [3] [7].

$$K_o = K_{mv} * K_{pv} * K_{cv} \tag{3-17}$$

where:

K_o = open loop gain (%/%)

K_{cv} = controlled variable (measurement) gain (%/EU)

K_{pv} = process variable (process) gain (EU/EU)

K_{mv} = manipulated variable (final element) gain (EU/%)

Note that in the literature on control, no analysis is usually given of the contribution of loop components to the open loop time constant or the open loop gain. In fact, everything outside of the controller is often referred to as the process, and there is no consideration or even appearance on a control diagram of final element and instrumentation details or location. Consequently, the open loop time constant, open loop gain, and the total loop dead time are called the *process time constant, process gain,* and *process dead time,* respectively. The total loop dead time is the summation of all pure dead times and time constants smaller than the second-largest time constant, no matter where they appear in the loop.

The open loop gain is commonly called a process gain despite the fact that it is dependent on the product of the final element, process, and measurement gains.

The open loop time constant is commonly called a process time constant even though it is the largest time constant in the final element, process, or measurement.

The total loop dead time is commonly called a process dead time even though it is really the summation of all time constants smaller than the second-largest time constant and the dead times in the final element, processes, and measurement.

Unified Approach

Since in columns, reactors, and vessels there are few interactions and small changes in controller outputs can cause resolution delays, smoothing the manipulated flow may be less important than the ability to control a primary loop at its set point. For the tightest control, the tuning is set to transfer as much variability from the process variable (controlled process output) to the controller output (manipulated process input) as possible. Suppose there is a Lambda factor of 0.1 and a dead-time-to-time-constant ratio of 0.1, as commonly found in temperature loops. In this case, the Lambda tuning equation reduces to ½ the open loop time constant divided by the product of the open loop gain and total loop dead time. The result is Equation 3-18, which is the simplified internal model control (SIMC) equation for tight control. Most documentation on Lambda tuning has focused on slowing down the response of fast loops so as to provide a smoother and a more consistent response, which is important for improving the coordination or reducing the interaction between loops on unagitated volumes. However, nothing prevents the Lambda tuning factor from being set less than 1.0 in order to provide much tighter control. Furthermore, unlike most other tuning knobs, setting the Lambda factor for more aggressive control does not cause the loop to become unstable. Rather, the loop just approaches the tuning equations cited for best load rejection capability. In fact, most of the equations developed in the 1960s and 1970s that focused on excellent load rejection have this common form of the controller gain, in which the gain is set equal to some factored ratio of the process time constant to the product of the open loop gain and total loop dead time [2]. Appendix C, "Unification of Controller Tuning Relationships," in this book, shows how Ziegler-Nichols, Lambda tuning, and internal model control tuning equations reduce to this common form. This reduction confirms diverse methods and opens the door for a unified approach to tuning [8].

$$K_c = 0.5 * \frac{\tau_o}{K_o * \theta_o} \tag{3-18}$$

For tight control, most of the tuning settings that have been developed are reducible to the common form, in which the controller gain is proportional to the time-constant-to-dead-time ratio.

The Lambda factor should be set based on the size of the open loop time constant and whether the goal is tighter control or better coordination and less interaction between loops. For cascade control, the secondary loop's time response should be five times faster than the primary loop's to prevent interaction between these loops [2] [3]. The ratio of the closed loop time constants of the primary to secondary loop should be greater than five to meet this criterion. If the Lambda factor for the secondary loop is 2, then the ratio should be large enough for even bench-top units. For a flow or speed loop that has a two-second time constant, the secondary loop

Lambda would be four seconds, which means the primary loop Lambda should be at least twenty seconds. This is not a problem for temperature, but could be a consideration for a pressure loop that is near the end of a batch when the off-gas flow is high.

Near Integrators

For "near" integrators, the integrator gain is the steady-state open loop gain divided by the open loop time constant, as shown in Equation 3-20. Note that the units of %/% per second or 1/second are better understood as the %/sec change in the ramp rate of the CV per percent change in the CO for the controller in manual. Since the lack of a steady state means that the process ramps when the controller is in manual, the integrator gain is measured as the change in ramp rates from before to after the change controller output. This method is also known as the "short cut method" and is an extension of the process reaction curve method developed by Ziegler and Nichols [3] [8] [9]. The dead time is the time from the change in controller output to a significant change in ramp rate. Normally, the output change is in the direction that will change the direction of the ramp, which means the dead time is the time until the reversal of slope [3].

$$K_i = \frac{K_o}{\tau_o} \tag{3-19}$$

$$K_i = \frac{CV_2/\Delta t - CV_1/\Delta t}{\Delta CO} \tag{3-20}$$

If we substitute Equation 3-19 into Equation 3-15, we have the controller gain for a "near integrating" process [2] [3] [8]. The tuning settings can now be determined from the change in ramp rates per % and total loop dead time and a desired closed loop time constant (λ). The tuning method or software does not need to wait for the time to steady state but can usually identify the new ramp rate in five loop dead times. If you consider that the loop dead time is 0.1 or less than the process time constant and that it would take five time constants plus the dead time if a steady state could be reached, then the test time to identity the model and hence the tuning settings is reduced by an order of magnitude. For example, the time required identifying the dynamics of a primary loop with a dead time of one minute and a time constant of ten minutes for a step change in controller output would be fifty minutes as a self-regulating response and five minutes as an integrating response. The actual time that software takes may be twice as long as stated to help screen out the effect of noise and disturbances. The test should be repeated in both directions if possible, and it may require performing tests at different batch times for several batches to see how much the dynamics change with batch conditions. For these reasons, the time saved by modeling processes with large time constants as "near integrators" is significant.

Modeling processes that have large time constants as "near integrators" can reduce by a factor of ten the open loop test time for identifying the process dynamics.

$$K_c = \frac{1}{K_i * (\lambda + \theta_o)} \tag{3-21}$$

If the controller on a process that has a large time constant is properly tuned with a Lambda factor of much less than one, then the time required for a closed loop test is reduced dramatically for the self-regulating model. This is because the closed loop time constant is a fraction of the open loop time constant.

Tight tuning of primary loops with large process time constants makes it possible to much more quickly identify self-regulating models for a closed loop test.

Loop Cycling

"Near integrating" and "pure integrating" processes develop slow rolling oscillations if the integral (reset) time is too small (i.e., too fast), and they have the unusual characteristic of an increasing propensity to develop these oscillations as the controller gain is decreased. In Appendix C, "Unification of Controller Tuning Relationships," Equation 3-22 is developed for a reset time that gives zero overshoot (critically damped response) [8]. If a slight overshoot and oscillations with a decay ratio of 1/20 is permissible, then the numerator can be reduced to 0.7 for faster load rejection. A range of 1 to 4 for the numerator is used in practice.

$$T_i > \frac{4}{K_i * K_c} \tag{3-22}$$

Integrating processes exhibit the following nonintuitive behavior: integral time is reduced if the controller gain is increased because the minimum integral (reset) time setting is inversely proportional to the controller gain.

Primary loops that have too large a controller gain develop oscillations that have a period of close to the ultimate period (e.g., four dead times). This is much faster than the slow rolling oscillations that are caused by combining a low controller gain and low reset time [2] [3]. Ultimate oscillations are uncommon in column, reactor, and vessel loops because the maximum gain (ultimate gain) that would trigger these oscillations is quite high for well-designed primary loops. Relatively fast oscillations in the primary loop often indicate a piping or mixing problem. More common if not hidden by historian data compression is the limit cycle from the resolution limits of the final elements and measurements (Figure 3-7), the slowly decaying oscillations caused by a reset setting too fast for a controller gain setting (Figure 3-8), or secondary loops that are tuned for too slow of a response (Figure 3-9).

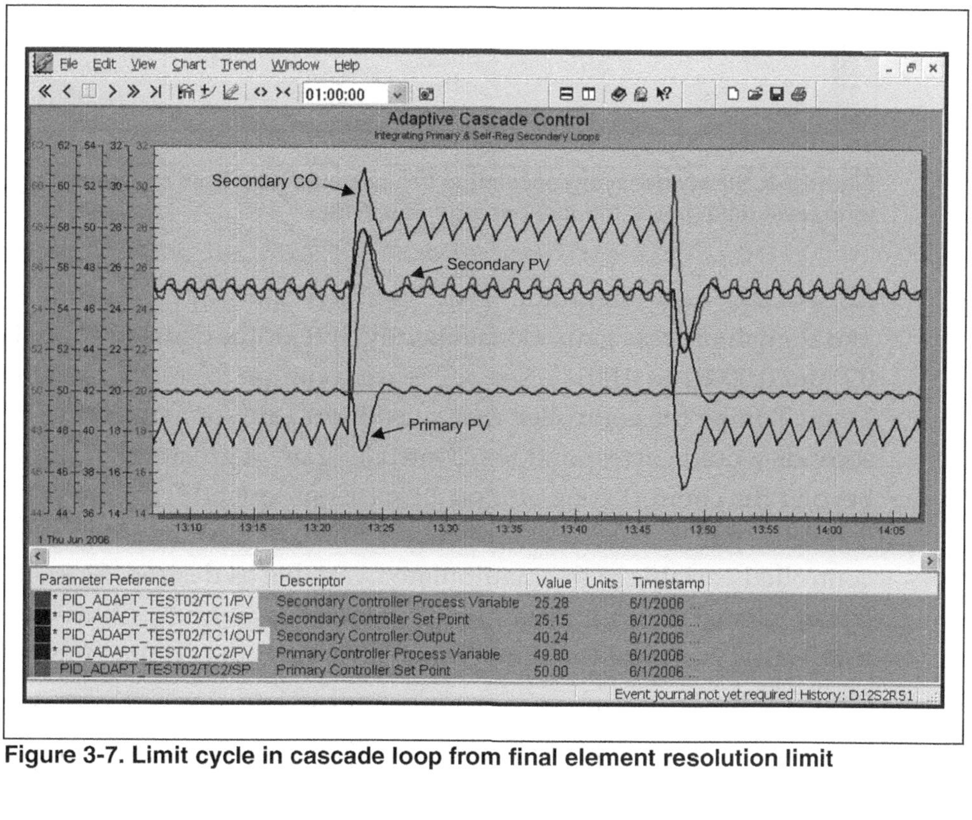

Figure 3-7. Limit cycle in cascade loop from final element resolution limit

Most of the oscillations in primary loops result from limit cycles caused by resolution limits, slowly decaying cycles from too much reset action, and interaction with secondary loops.

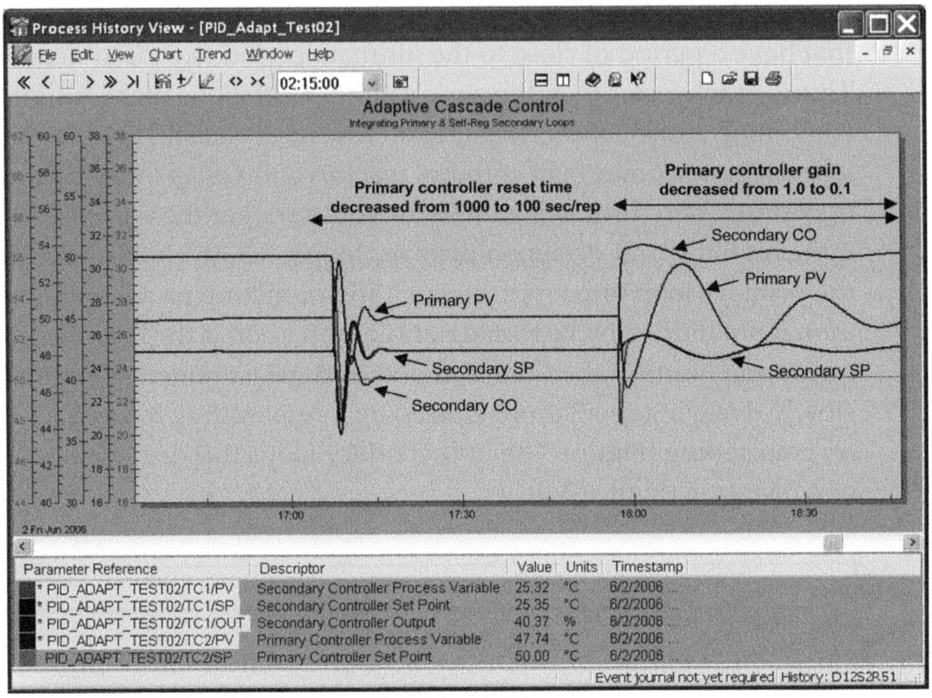

Figure 3-8. Slowly decaying oscillation in a cascade loop from an integrating primary loop reset time that is too fast per equation 3-13j

Most of the "near integrator" processes have a large process time constant and a small process gain. Consequently, integrator gains of about 0.001 to 0.1 and 0.00001 to 0.001%/sec per % are common for reactor temperature loops. Tuning the controller with maximum gain for a dead time of 100 seconds would correspond to a controller gain of greater than 10, which is beyond the comfort zone of most users. Consequently, a more moderate controller gain is used that transfers as much of the variability from the controlled variable to the manipulated variable as desired. Since this controller gain is often significantly less than the maximum allowed by Equation 3-21, it is critical that Equation 3-22 be used to prevent slow rolling oscillations.

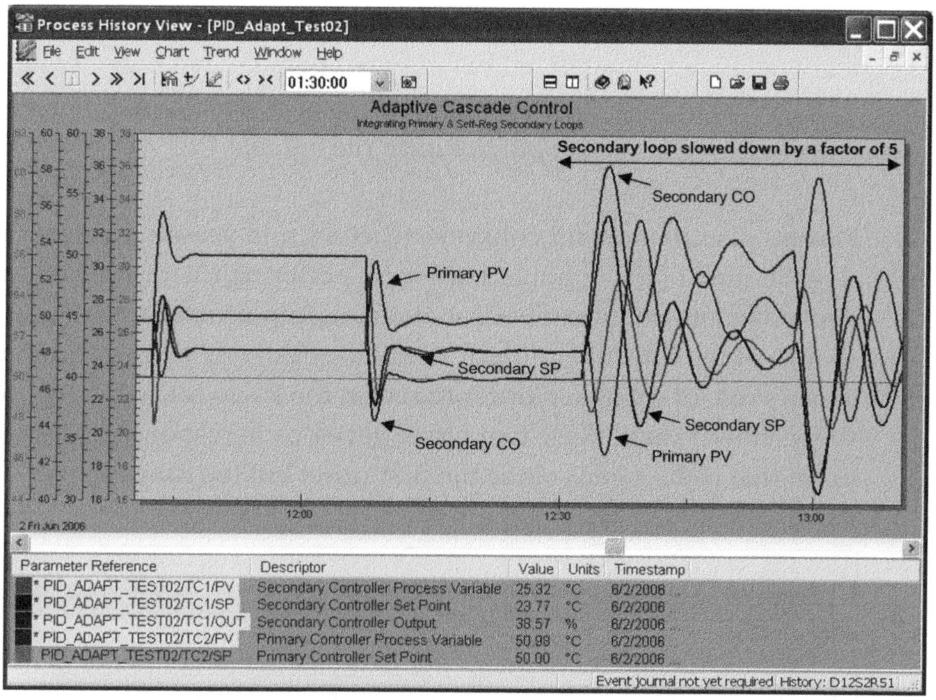

Figure 3-9. Interacting oscillations in cascade loop from a slow secondary loop

Since the primary loop controller gain is set significantly below its maximum, it is imperative that the reset time be increased to prevent slowly decaying oscillations.

It is important to remember that there is a balance point with all integrators, that the degree of imbalance determines the ramp rate, and that the ramp does not change direction until the unbalance changes sign. The implications of this behavior are explored in Section 3-5 on optimizing the set point response.

The ramp rate of an integrating process is proportional to the magnitude of the difference between the feed and the demand or exit flow, and the ramp rate cannot reverse direction until the unbalance reverses sign.

The ramp rate of the integrator response is inversely proportional to the mass holdup, which is the volume for a constant density. Of course, the scaling of flows should match the volume, but the sizing and unbalance may not match. Bench-top units tend to have much faster ramp rates. The change in ramp rate (integrating process gain) must be considered when moving tuning settings to production units.

The ramp rate of an integrating process is inversely proportional to the volume.

Process disturbances for columns, reactors, and vessels are generally very slow because of the volume. The loop's performance for slow disturbances depends more on integral action than on proportional action. The rate of change of the controller output needs to effectively exceed the load disturbance's rate of change in order to return the PV to set point. Figure 3-10 shows how a small persistent offset develops in response to a second upset that is the same size as the first upset but ten times slower.

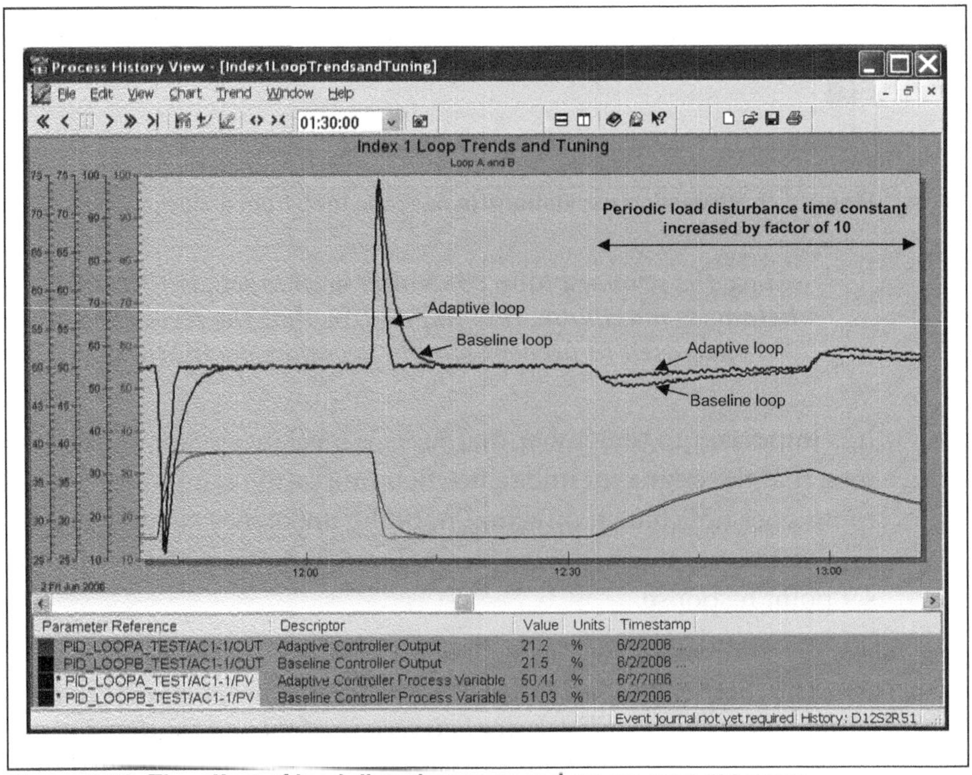

Figure 3-10. The effect of load disturbance speed on process recovery

Lambda tuning equations specifically designed for pure integrating processes are available. Appendix C, "Unification of Controller Tuning Relationships," details the use and reduction of these more exact equations into the simpler unified form shown here for the "near" integrating processes of common column, reactor, and vessel loops. The appendix shows how the more exact equations for "pure" integrating processes provide a maximum controller gain that is about 50 percent larger than what is estimated from Equation 3-21 for the low integrator gains seen in columns, reactors, and vessels. Since the actual controller gain used in practice is typically far below this maximum, the more complex equations in Appendix C are more useful for deploying in software packages for controller tuning than for understanding or estimating tuning settings.

3-4. Adaptive Control

Theory and Reality

Process control systems assume a constant linear process. Unfortunately, all process variables and control valves are nonlinear to some degree: the process response to a given change in the controller output changes with batch time. The lack of consistency in the process response has significant implications for the process's performance not only in terms of tuning controllers but of recognizing degradations and achieving optimums [12].

Road Maps and Terrain

Consciously or subconsciously, tuning controllers involves a tradeoff between performance and robustness. The controller's ability to tightly control at an operating point is inversely proportional to its ability to weather changes in the plant's behavior without become oscillatory. The operating environment for most loops is stormy, and the last thing you want is for a control loop to introduce more variability. Consequently, all controllers are detuned (backed off from maximum performance) to some degree to provide a smooth response, despite the inevitable changes in the process dynamics. A PID controller approaches turns cautiously since it doesn't know what lies ahead [12].

PID controllers are backed off from best performance because of the uncertainty of tuning settings.

Controller tuning settings can be computed from a first-order or an integrating-plus-dead-time process model. The changes in the model parameters reveal changes in the cells, process conditions, equipment, final elements, and sensors. The size, direction, and characteristics of these changes can provide a road map and knowledge of the terrain [12].

Changes in the parameters of a dynamic process model identified by an adaptive controller provide insight into changes in the process, equipment, and sensor.

Glimpses and Grimaces

Nearly all of the industrial adaptive controllers presently used in industrial processes require that changes in the process variable be observed over rather a long time and they show the results in terms of new tuning settings. The tuning rules are imbedded and usually unknown. The most commonly used adaptive controller today operates by using pattern recognition and, if it's deemed necessary, it increases the controller gain to induce oscillations. The size of the transients or oscillations and the time required for identification can translate into significant process variability and an adaptation rate that is slower than the rate of change of the process parameters. In fact, most adaptive controllers are playing catch-up even if they have seen the same situation a thousand times before. At best, these controllers provide a snapshot of the current tuning requirements but no real process insight into where the process has been or where it is going. Also, sudden unexplained shifts in the tuning settings or bursts of oscillations reduce the operator's confidence and decrease the likelihood that the controllers will run in the adaptive mode and be used in future applications [11] [12].

Model-free and pattern-recognition adaptive controllers do not offer process knowledge and are playing catch-up.

Watching but Not Waiting

The next generation of adaptive controllers can identify a process model quickly and automatically and provide process model parameters that can be displayed, trended, and diagnosed. Furthermore, these controllers remember the results for similar conditions, eliminate repetitious identification, and take the initiative [12].

An adaptive controller with these desirable features has been demonstrated in plant tests. The controller can identify the dead time, process gain, and time constant for both manipulated and disturbance variables and save these as a function of a key variable. The user can use the recommended tuning method or elect to choose an alternative method to compute the current tuning settings for the current and memorized conditions. When the key variable indicates that the process has changed, the tuning is then scheduled based on the process model saved in the operating region. The adaptive controller remembers the results from previous excursions and does not wait to recognize old territory. For example, in loops that have nonlinear installed valve characteristics and nonlinear controlled variables, the model and tuning are scheduled based on the controller output and input, respectively. To change dynamics as the batch progresses, the model and tuning are scheduled based on totalized feed. The adaptive controller takes preemptive action based on operating region and uses the opportunity to refine its knowledge of the process model. Changes in these models can flag changes in seed cultures [12].

Adaptive controllers should learn, remember, and utilize knowledge gained from previous batches.

The adaptive controller computes the integrated squared error (ISE) between the model and the process output for changes in each of three model parameters from the last best value. To explore all combinations of three values (low, middle, and high) for three parameters, twenty-seven models are ultimately generated. The correction in each model parameter is interpolated by applying weighting factors that are based on the ISE for each model, normalized to a total ISE for all the models over the period of interest. After the best values are computed for each parameter, they are assigned as the middle values for the next iteration [13]. This model switching with interpolation and recentering has been proved mathematically by the University of California, Santa Barbara, to be equivalent to least square identification and provides an optimum approach to the correct model [14]. The search is actually done sequentially, first for the process gain, then the dead time, and finally the time constant, which reduces the number of models to nine [13]. Figure 3-11 shows the setup of an adaptive controller that identifies the process model for the controlled variable's (ΔCV) response to changes in the controller output (ΔCO) and disturbance variable (ΔDV). This response is then used to compute the feedback controller tuning settings and the feedforward dynamic compensation, respectively [12].

The adaptive controller starts in the "Observe" mode, in which it continuously and automatically identifies the process model when it sees changes in the controller's set point, output, or feedforward. The adaptive controller can also be switched to the "Learn" mode, in which it updates the feedback and feedforward process models in each region for which it sees an

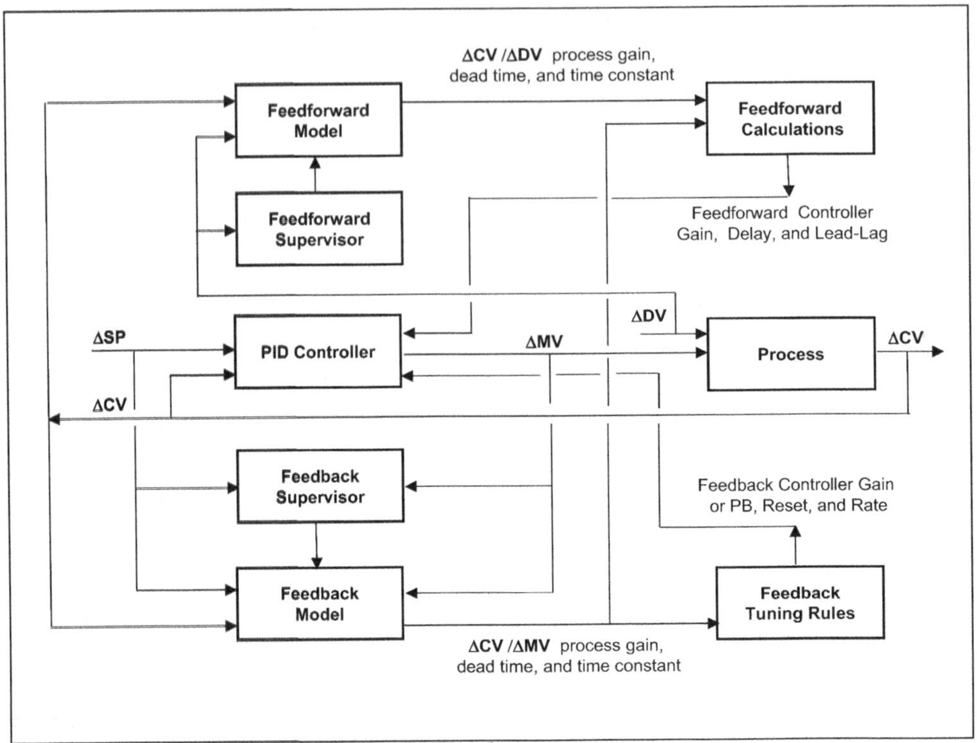

Figure 3-11. Adaptive controller setup based on identifying process models

excursion. The next option is the "Schedule" mode, where the adaptive controller uses the models in each region to change the controller settings. The highest level is the "Adapt" mode, in which the adaptive controller immediately uses any identified improvements [12].

An adaptive controller can be run in the "Observe" and "Learn" mode in order to see what it can capture as process knowledge as a function of batch time or variables.

Back to the Future

This new generation of adaptive controllers allows all PID loops to run in the adaptive mode and provides process model parameters that are saved in a data historian and analyzed for changes in the plant, sensors, and valves. The information on changes in the process model may be directly used to monitor loop performance and to provide more intelligent diagnostics. The models can provide the dynamics for simulations and identify candidates for advanced control techniques. For example, loops that have large dead times or a one-way integrating response are prime candidates for model predictive control [12].

3-5. Set-Point Response Optimization

Nothing Says Forever Like Tradition
There are four major practices for starting up a loop that has a large process time constant or slow ramp time compared to the dead time [15]:

Loop Practices for Fast Batch and Startup Response

1. Switch controller to auto with the final set point.
2. Switch controller to auto with an initial set point and then switch to final set point.
3. Put controller in manual or output tracking with final set point, set valve to its normal position, wait, and switch to auto.
4. Put controller in manual or output tracking with final set point, set valve to extreme position, wait, switch valve to normal position, wait, and switch to auto.

Figure 3-12 shows the batch or startup response of a pressure loop that has an integrating response for practices 1, 2, and 4. Practice 3 is not shown because it is not viable for integrating processes. Other practices exist, such as ramping the set point, for unit operations in which it is desirable that the process variable's approach to set point and the output to its final resting value are moderated or that a profile is enforced [15].

The controller output must be positioned beyond the final resting value (balance point) in an integrating process in order to get the CV to move toward set point.

In the first and second practices, the controller output is at its initial value at one end or the other of the output scale (often zero). All methods assume that the pump and block valves have already been started and opened, respectively [15].

In the first practice, if the loop is tuned to minimize variability in the controller output, which is the case for surge volume level control, then the batch phase may time out before the process reaches set point. For example, if the process time constant is 50 minutes and a Lambda factor of five is used, then the closed loop time constant is 250 minutes and the time to

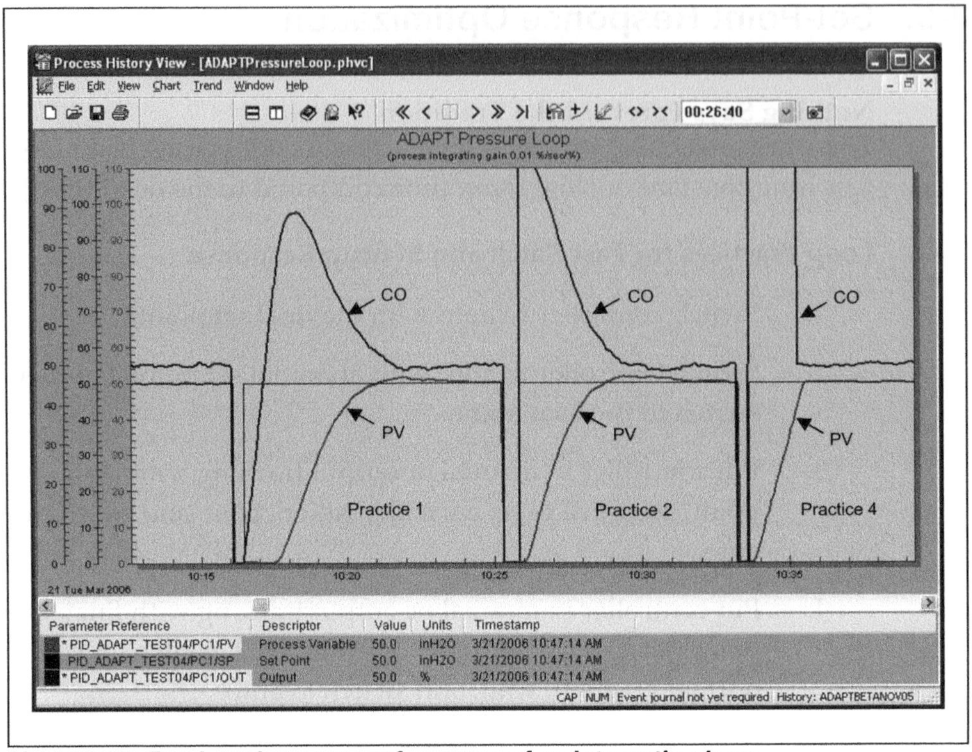

Figure 3-12. Batch and startup performance of an integrating loop

reach 98 percent of set point is 1000 minutes (four closed loop time constants). A similar situation exists for slow ramp rates [15].

A dead time that is much faster than the process time constant or ramp rate usually means that a Lambda factor of less than one (i.e., a closed loop time constant or arrest time that is less than the open loop time constant or arrest time) is permissible to achieve stability and is desirable to achieve fast control of these process variables. This is particularly important for the practice 1 because you are relying on reset action to get you set point. All the batch or startup responses in Figure 3-12 use a Lambda factor of less than one [15].

If the kick from the proportional mode is negligible, the burden is entirely on the integral mode (reset action), and the approach to set point will be much slower.

In the second practice, the set point is changed from its initial to final value one execution or more after the controller is switched to auto. Note that if you switch the set point within the same execution of the module as the switch of the mode, you will probably end up with the same response as the first practice. In the second batch or startup response, the set point change kicks the output, which gives the process variable a boost on its way to set point. The time to reach set point (rise time) is nearly cut in half, but the settling time is about the same. Since the overshoot is minimal, the rise time might be more important. Also, the controller tuning could be tweaked to reduce settling time [15].

If the set point change is made during the same execution as a mode change, there will probably be no kick from the set point change, despite a β factor greater than zero.

Many of the more astute automation engineers pre-position the controller output by what is called a *head start* or *process action*. For self-regulating loops, the valve position might be set at what was considered to be a normal throttle position or final resting value (FRV), as seen from previous trends when the process variable had settled out at set point. This corresponds to a Lambda factor of one because, if held at this position, it drives the process variable with a time constant that is equal to the process time constant [15].

A Lambda factor of one for self-regulating loops gives an approach to set point at the speed seen in the open loop response for a controller output that is moved to the FRV.

For integrating responses, the process variable won't go anywhere until the valve is positioned beyond its FRV. This leads us to practice 4, in which the valve is set to an extreme position allowable by the process in order to give the fastest approach to set point. Then the brakes are slammed on so the process variable does not run over the set point. The question is, When do you hit the brakes [15]?

The Wait

The plot for the fourth practice shows the response for a technique in which the rate of change is computed from the change in the controlled variable (CV) over a time period long enough to get a good signal-to-noise ratio. The old value of the CV, created by passing the CV through a dead-time block, is subtracted from the new CV. The delta CV is divided by the block dead time to create a rate of change. The rate of change multiplied by the process dead time is then the predicted change in the CV that when added to the new CV is the predicted end point as shown in Equation 3-23. When the end point equals or exceeds the final set point, the controller output is switched from maximum throttle to its FRV. It is held at this FRV for one process dead time and is then released for feedback control. This method compensates for nonlinearities and disturbances that are evident at the time to hit the brakes [15].

The controlled variable's rate of change multiplied by the loop's dead time and added to the old controller's output provides a prediction of the end point.

$$CV_f = [(CV_n - CV_o) / DT] * \theta_o + CV_n \tag{3-23}$$

where:

CV_f = predicted CV one dead time into the future (%)

CV_n = new CV (%)

CV_o = old CV (output of dead time block) (%)

DT = DT block dead time (sec)

θ_o = total loop dead time (sec)

If the process dead time is underestimated, the loop overshoots the set point. Therefore, it is important to be generous in the dead time estimate. It is especially important that the dead time not be too short for the zero load integrating process, where the FRV is zero and there is nothing to bring the process variable back to set point [15].

It is better to overestimate the dead time when predicting the end point.

Without Dead Time I Would Be Out of a Job

If the loop dead time is zero, the loop could switch to the FRV when the CV reached set point. Furthermore, the sky is the limit, for the controller gain and feedback action could provide instantaneous correction. My lifestyle is largely the result of dead time.

A better term than process dead time is *total loop dead time* because there are many sources of dead time outside of the process. The biggest source for slow ramp times and process time constants is the measurement and valve resolution. The time it takes for the change to get through the resolution limit is dead time. The dead time from the measurement and valve resolution are inversely proportional to the rate of change of the process variable and controller output, respectively. Consequently, for closed loop tests the identified dead time depends on the controller tuning and the size of the change in the set point. Fortunately, the changes in valve position are quite large, and the dead time from resolution is minimal for the optimal switching method described in practice 4.

Other sources of instrument dead time include measurement transportation delay, sensor lags, transmitter dampening, analog and digital filters, module execution time, valve dead band, and actuator prestroke dead time.

An adaptive controller can identify the total loop dead time accurately if the trigger points in terms of output changes are large enough. Note that the ultimate proof in the pudding is the output change rather than the set point change. This is because output change includes the effect of tuning and is ultimately what is driving the process. Given the measurement and valve resolution, the adaptive controller with its knowledge of the integrating process gain can correct the observed dead time to give a value that is closer to the output changes associated with the optimal switching [15].

An adaptive controller can also identify the integrating process gain. This can be used with the current ramp rate and the pre-positioned extreme controller output to estimate the FRV, following Equation 3-24. Note that if the extreme output (CO_x) is less than the FRV, the signs of each expression are reversed to get a positive FRV. Of course, limits should be enforced on the calculated value, and it may be desirable to estimate the new FRV by using a portion of the difference between the calculated FRV and the last captured FRV added to the last captured FRV. For primary loops in a cascade control system, the extreme output must match up with the set point limits of the secondary loop and the FRV *is* a set point of the secondary loop. It is necessary to keep the units of the process variable and output consistent with the process integrating gain. If process integrating gain is expressed in %/sec/%, then the process variable and output must both be in % [15].

An adaptive controller can identify the process gain and dead time that are essential for improving the evaluation, performance, and monitoring of optimal switching.

For integrating processes and $CO_x > FRV$:

$$FRV = CO_x - [(CV_n - CV_o) / DT] / K_i \tag{3-24}$$

where:

FRV = final resting value (%)

K_i = integrating process gain (%/sec/%)

CO_x = controller output at extreme allowed by process (%)

CV_n = new CV (%)

CV_o = old CV (output of dead time block) (%)

DT = DT block dead time (sec)

With a little ingenuity, similar equations can be developed for estimating the FRV of self-regulating processes based on an identified process gain. These equations can be put on line in the observation mode to see how well they estimate the FRV before you actually use the FRV for practice 4. If the FRV is too variable and cannot be accurately captured or calculated, it is best to revert to the second practice. The second practice depends more heavily on the controller's tuning and in particular on the relative amount of proportional and reset action. This is because the tuning is responsible not just for correcting the FRV but for taking the output all the way from its extreme to the FRV. It is important that gain dominates reset action in the approach to set point. Proportional action must kick the output to the allowable extreme and then back it off as the CV approaches set point. This is despite the effect of the reset action, which works to force the output to its limit until the CV crosses set point [15].

The optimal switching technique based on the CV's rate of change is ideally suited for an integrating or ramping process. However, it works well for self-regulating processes in which the fastest possible approach to set point is desired. It also reduces the dependency on tuning since the PID only has to correct for errors in the dead time and the FRV [15].

Exercises

3-1. Why does a valve set at a fixed position set by operations or sequences to deliver a flow on a process flow diagram result in excessive process variability?

3-2. Why are controller gains for vessel temperature control not as high as permitted by the large process time constant to dead time ratio?

3-3. Why do PID structures with proportional mode on error provide a faster set point response?

3-4. For the ISA standard form of the PID, what is the limitation on the rate time setting?

3-5. How does the dynamic reset limit improve loop stability?

3-6. Why does the open loop response of a vessel temperature loop ramp like an integrator?

3-7. To reduce the slow rolling oscillations in an integrating process from too low of a reset time setting, what should be done to the gain setting?

3-8. What are the advantages of software that identifies the process dynamics for various modes and ranges of operation?

3-9. For a slow integrating type of response, what can be done to make the set point response as fast as possible?

References

1. McMillan, Gregory K., *A Funny Thing Happened on the Way to the Control Room*. Reprint via ProQuest UMI "Books on Demand", 1989.

2. McMillan, Gregory K., *Good Tuning: A Pocket Guide*. 2d ed. ISA, 2005.

3. McMillan, Gregory K., *Tuning and Control Loop Performance*, 3d ed. ISA, 1992.

4. Åstrom, Karl, and Hägglund, Tore, *Advanced PID Control*. ISA, 2006.

5. McMillan, Gregory K., and Weiner, Stan, "Control Mythology." *Control* (April 2006).

6. Shinskey, F. G., *Feedback Controllers for the Process Industries*. McGraw-Hill, 1994.

7. Blevins, Terrence L., McMillan, Gregory K., Wojsznis, Willy K., and Brown, Michael W., *Advanced Control Unleashed: Plant Performance Management for Optimum Benefits*. ISA, 2003.

8. Skogestad, Sigurd, "Simple Rules for Model Reduction and PID Controller Tuning." *Journal of Process Control* 13 (2003): 291-309.

9. Ziegler, J. G., and Nichols, N. B., "Optimal Settings for Automatic Controllers." *Transactions ASME* 64, no. 11 (1942):759.

10. Wilson, Grant, McMillan, Gregory K., and Boudreau, Michael, "PAT Benefits from the Modeling and Advanced Control of Bioreactors." *Emerson Exchange*, 2005.

11. Trevathan, Vernon L. (editor), *A Guide to the Automation Body of Knowledge*. "Chapter 3 - Continuous Control." Wade, Harold (author), ISA, 2005.

12. McMillan, Gregory K., Sowell, Mark, and Wojsznis, Peter, "The Next Generation – Adaptive Control Takes a Leap Forward." *Chemical Processing*, September 2004.

13. Hespanha, Joao P., and Seborg, Dale E., "Theoretical Analysis of a Class of Multiple Model Interpolation Controllers." Presentation at AIChE Conference, San Francisco, 2003.

14. Wojsznis, Willy K., Blevins, Terrence L., and Wojsznis, Peter. "Adaptive Feedback/Feedforward PID Controller." Presentation at ISA EXPO, Houston 2003.

15. McMillan, Gregory K., "Full Throttle Batch and Startup Response." *Control*, May 2006.

16. Shinskey, F. Greg, "The Power of External Reset Feedback." *Control*, May 2006.

4
Process Dynamics

Learning Objectives

A. Understand how dead time sets the ultimate performance limit and maximum possible aggressiveness of the controller tuning.

B. Learn how the design of equipment and automation systems affect dead time.

C. Become familiar with ways to estimate loop performance from tuning settings.

D. Be aware of how limit cycles originate from the final element resolution.

E. Learn the three basic parameters used to define the process response.

F. Recognize the differences between self-regulating and integrating responses.

G. Be able to convert self-regulating response parameters into an integrating parameter.

4-1. Introduction

The dynamics of a process response can be modeled by three types of parameters: dead time, time constant, and process gain. The dynamics are identified from the plant's operation, and they determine both the process's ultimate performance and the actual performance of the control system. The process dynamics are used to compute PID controller settings. They could also be employed directly to predict the trajectory of process outputs from past process inputs in future applications of model predictive control (MPC) for columns, reactors, and vessels. Although MPC technology has been used almost exclusively on continuous processes, recent batch applications are promising, and it is anticipated that column, reactor, and vessel applications are possible.

Section 4-2 of this chapter discusses the largest sources of dead time in the automation and process systems, the impact of total dead time on the controller's capability and tuning, and the procedure for estimating the performance based on the tuning settings. Section 4-3 details the characteristics of a self-regulating process response, the way small time constants effectively become dead time, and the reasons why the location of the largest time constant affects control. Section 4-4 delves into the integrating process response of the more important process variables as well as the implications of this response for process testing and controller tuning,

4-2. Performance Limits

Impact of Loop Dead Time

Control loop dead time is the period of time after a change in the controller output (manipulated variable) in which there is no recognizable change in the process variable (controlled variable). For feedback control, the controller output must change a process input in order to cause a change in a process output that is measured and seen by the controller. The performance of the control loop is ultimately limited by the loop dead time, which is the time it takes to make and see a change. However, the actual performance is determined by the controller tuning. The dead time sets how aggressively a controller can be tuned without becoming unstable. In practice, there is always a trade-off between performance and robustness (that is, the controller's relative ability to be stable despite changes in process dynamics). In the process industry, controllers are generally tuned with settings that are far from the performance limit allowed by process dynamics [7]. This is particularly true for the primary loops on columns, reactors, and vessels because the process variable's rate of change as determined by a large process time constant or a small integrating process gain is slow compared to the size of the process dead time.

The ultimate limit to controller performance is determined by the loop dead time, but the actual limit is usually set by the controller tuning.

The control loop must respond to changes in its set point and loads (disturbances). The disturbances can originate from changes to any of the process inputs, such as flows, charges, and intracellular metabolism and composition. Set point changes to primary loops (temperature targets) are rather minimal, although the optimum temperature may change. However, the set points to the secondary loops (changes in gas or reagent flow, agitation, and coolant demand) are continual and cover a wide range. This is so they can deal with the one to three order-of-magnitude changes during the batch [1].

The control valve's resolution and dead band should be determined per ISA standards ISA-TR75.25.01-2000 (R2006) and ISA-TR75.25.02-2000 (R2006) on valve-response testing and measurement. Although it is difficult to generalize, sliding stem valves that have a packless or low-friction design with digital positioners have a resolution limit of 0.05 percent to 0.1 percent and a dead band of 0.1 percent to 0.2 percent. Rotary control valves tend to have a larger resolution limit and dead band. Control valves without positioners have an order-of-magnitude or larger resolution limit and dead band [3] [4] [5].

For the small actuators used on valves for column, reactor, and vessel control, the prestroke dead time and stroking lag times associated with pressure changes in the actuator are negligible. The largest source of dead time in a valve response is the resolution limit and dead band whenever the rate of change is slow in the controller output. The dead time can be estimated as the resolution limit or dead band for changes in the same direction and opposite direction, respectively, divided by the rate of change of the controller output [3] [4] [5]. However, as the total size of the change approaches the resolution limit, the dead time gets much larger, particularly in pneumatic positioners [5]. Some digital positioners can be tuned to minimize this increase in dead time. The larger and faster changes in the controller's output from less sluggish tuning settings can minimize this dead time.

The control valve dead time from resolution and dead band can be minimized by using a digital positioner and faster controller tuning settings.

The start of a change in flow as a result of a change in speed is almost instantaneous unless a dead band has been configured in the variable speed drive (VSD) to prevent a response to noise. The rate of change of speed might be limited in the VSD for motor load, but the rate limiting is usually fast enough so that it is not a consideration.

In smaller-scale bench-top and pilot plant–size columns, reactors, and vessels, there is significant use of peristaltic pumps for liquid additions such as base. This is because it is easy to take the tubing out of the pump roller bar mechanisms and clean and sterilize the tubing associated with these feeds [1].

Impact of Controller Tuning Settings

In a single-loop configuration, the output of the process controller goes directly to a VSD or a control valve. In a cascade loop, the output of the primary process controller becomes the set point of a secondary controller. The most common secondary controller is the flow or speed controller. These secondary loops compensate for load changes and disturbances before they affect the primary process controller. However, the secondary controller response must be five times faster than the primary controller response, or interaction will develop between the two controllers. The secondary controller must be tuned for a fast response. For flow and speed control, this corresponds to a reset time of less than six seconds. The performance of a cascade control system that has a slow secondary loop may be worse than if the primary controller output would set the VSD's variable frequency directly without tachometer feedback.

The secondary controller must be tuned to be five times faster than the primary controller or cascade control can do more harm than good.

According to the ISA and Fieldbus standards, the tuning settings for a proportional-integral-derivative (PID) controller are gain (K_c), reset time (T_i), and rate time (T_d). The gain setting is dimensionless, and the time settings are in seconds per repeat for reset and in seconds for rate. For reset time, the setting may be listed in terms of seconds with the seconds per repeat implied. In some controllers, proportional band in percent ($100\%/K_c$) is used instead of gain, and reset action in repeats per minute ($1/(T_i/60)$) is used instead of a reset time in seconds per repeat. Since these settings are inversely related to the ISA settings, it is critical that users understand what type of tuning settings is used as well as their units. In most cases, derivative action is not used (rate time is zero), which results in a proportional-integral (PI) controller. The effect of PID form and structure on performance is discussed in Chapter 3, on basic feedback control.

Another important consideration is that users set the controller action so it is the opposite of the process's action if any reverse action of a final element is accounted for by a signal reversal. This reversal can be achieved in field devices, but it is preferable that it be done for maintainability and visibility in the blocks in the configuration, such as the "splitter" and the "analog output" blocks. Barring any improper signal reversals, the controller action is "reverse" if the process action is "direct" and vice versa for a single manipulated variable. For a "reverse" action controller, an increase in the process variable above set point decreases the controller output. A reverse-acting (fail-open or increase-to-close) final element is usually not a concern except perhaps for a coolant or vent control valve.

Signal reversals typically occur when the controller output sets multiple flows (i.e., multiple manipulated variables) by either split-ranged or simultaneous action. The signal reversal is normally done in "splitter" and "analog output" (AO) blocks for split-ranged and simultaneous action, respectively. The output action can also be reversed in the controller block, but this practice is not recommended unless the PID is directly addressed and there is no "splitter" and AO block.

If the conversion between gain and proportional band and between reset action and reset time or controller action is wrong, nothing else matters.

Equation 4-1 for integrated error is derived, as shown in Appendix D, from the response of a standard form of the PI controller to a load upset where the open loop error (E_o) is the magnitude of the error if the controller is in manual. The integrated error (E_i) is the total area between the process variable and the set point for the controller in automatic [3] [4]. The equation shows that the integrated error is inversely related to controller gain and directly proportional to reset time.

$$E_i = \frac{1}{K_o * K_c} * T_i * E_o \tag{4-1}$$

where:

E_i = integrated error (% seconds)

E_o = open loop error (%)

K_c = controller gain

K_o = open loop gain (also known as process gain) (%/%)

T_i = controller reset time (seconds)

If we take the ratio of the integrated errors at a constant open loop gain for new and original tuning settings and subtract this ratio from 1, the open loop error cancels out. We then end up with the expression within the parentheses in Equation 4-2, which gives an estimate of the fractional improvement in integrated error for load changes achieved by more aggressive tuning. If we multiply this fraction improvement by the control error, the result is the possible improvement in the standard deviation as a result of better tuning. The equation assumes that the loop is stable. If the loop is oscillating, it might be better to have less aggressive tuning, but this case is rare except at the start of the batch when demand is extremely low. Equation 4-2 does not address the portion of the standard deviation that is caused by limit cycles or noise since these are set by the devices, equipment, and process. Though Equation 4-1 was derived for load changes, it also gives a relative indication of the performance improvement for set point changes. For cascade control, Equation 4-1 should first be used to improve the tuning of the secondary loop before it is applied to the primary loop.

$$\Delta S_c \cong Maximum\left[\left(1.0 - \frac{K_{co}}{K_{cn}} * \frac{T_{in}}{T_{io}}\right), 0.0\right] * \Delta E_c \tag{4-2}$$

where:

ΔE_c = control error caused by load or set point changes (%)

K_{cn} = new controller gain

K_{co} = old controller gain

ΔS_c = improvement in standard deviation from new controller tuning (%)

T_{in} = new controller reset time (seconds)

T_{io} = old controller reset time (seconds)

The control error (ΔE_c) that is caused by load or set point changes can be estimated from the mean absolute error (MAE_c) over a test period. The current test time is the module's execution time added to the last value of the test time multiplied by a fractional "forgetting factor" determined by the supervision time, as detailed in Equation 4-3. Similarly, the integrated absolute error is the absolute error added to a last value of the integrated absolute error forgotten to a degree again determined by the supervision time, as detailed in Equation 4-4. The mean absolute error is just the integrated absolute error divided by the test time, as shown in Equation 4-5. Finally, the control error (ΔE_c) for use in Equation 4-2 is computed, per Equation 4-6, as the absolute difference between the mean absolute error and the standard deviation of the capability of the process (S_{cap}) set by noise. This can be estimated from the moving rate average of successive differences of the deviation from set point, as illustrated by Equations 4-7 and 4-8. The supervision time is typically either the duration of an operator shift or a batch phase. If the configuration is not set up so the set point tracks the process variable when the PID control algorithm is in the manual or remote output modes, then the error should be zeroed out until the controller is actually in service (automatic, cascade, or remote cascade modes).

$$T_n = T_e + (1 - 1/T_s) * T_{n-1} \tag{4-3}$$

$$IAE_n \cong | CV_n - SP_n | + (1 - 1/T_s) * IAE_{n-1} \tag{4-4}$$

$$MAE_c \cong IAE_c / T_n \tag{4-5}$$

$$\Delta E_c \cong | MAE_c - S_{cap} | \tag{4-6}$$

$$MR_n = \mathrm{r} * [| CV_n - SP_n | - | CV_{n-1} - SP_{n-1} |] + (1 - r) * MR_{n-1} \tag{4-7}$$

$$S_{cap} \cong MR_n / 1.128 \tag{4-8}$$

where:

CV_n = controlled variable for current scan (%)

CV_{n-1}= controlled variable for last scan (%)

IAE_n = integrated absolute error for test time (% sec)

MAE_n= mean absolute error for test time (%)

MR_n = moving rate average at scan n (%)

r = weighting factor (0.02)

SP_n = set point for current scan (%)

SP_{n-1} = set point for last scan (%)

S_{cap} = standard deviation of the capability of process as determined by noise (%)

T_e = module execution time (seconds)

T_s = process supervision time (seconds)

T_n = current test time (seconds)

An adaptive controller has been developed that identifies the process model and the new controller tuning settings for Equation 4-2. This makes it possible to estimate the possible improvement in the standard deviation through better tuning [5] [6] [7].

The improvement in control loop performance can be estimated from the change in tuning settings and the mean absolute error.

The controller tuning's dependence on process dynamics, including dead time, is detailed in Chapter 3. It is important to remember that while the most aggressive tuning settings and the ultimate performance depends on the relative size of the dead time, the actual performance still depends on the tuning settings used. In other words, if a controller is detuned, it will perform the same as a controller with more dead time.

Sources of Dead Time

The purpose of a control system is to recognize and deal with change. Figure 4-1 shows the path from a change in controller output to a change in the process variable. A dead time or a time lag is associated with each block in the path. The total loop dead time can be estimated as the sum of the time delays and time lags of each block [3] [4] [5]. It can be visualized as the total delay required for a signal to make one cycle around the loop. The starting point of the change can be anywhere in the loop.

The controller has a dead time that is, on average, one-half of the module execution time, which is typically one to five seconds. Excluding resolution limits, the variable speed drive's dead time is negligible unless a dead band has been added for noise rejection.

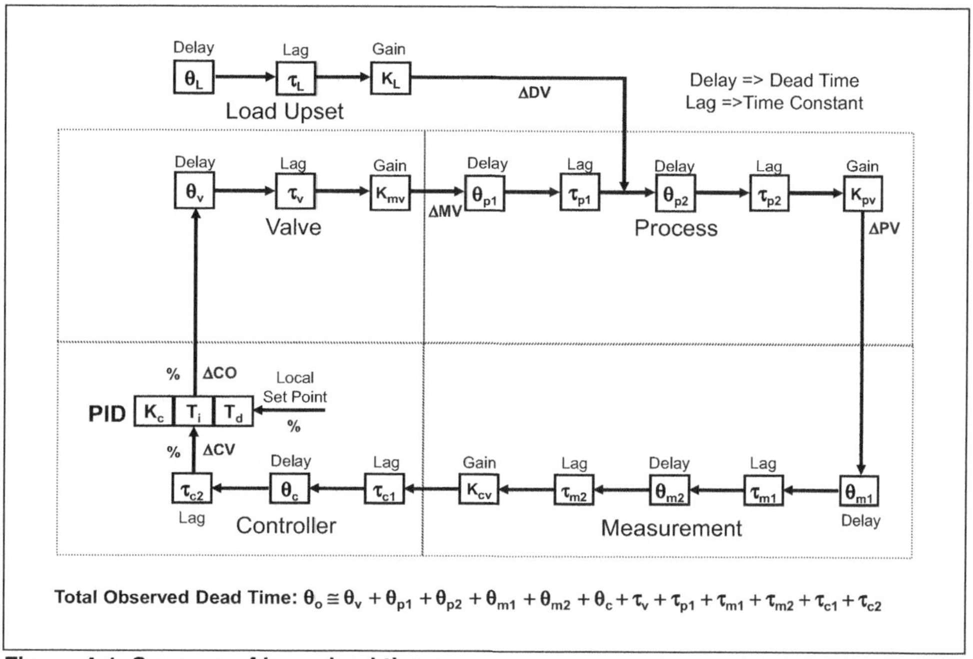

Figure 4-1. Sources of loop dead times

If the sensor is too close to the wall or a baffle, the lower fluid velocity translates into a slower dispersion and a slower inherent response, which is greatly aggravated by an increased propensity for fouling [6]. In the extreme case of stagnation, the sensor does not see a process that is representative of the mixture. If the sensor was located externally, there would also be a transportation delay associated with the piping from the column, reactor, or vessel to the sensor.

Slow fluid velocities increase the fouling and time lag.

The process dead time caused by liquid mixing (ignoring transportation, injection, stagnation, and sensor delays) for a well-mixed vessel with axial agitation can be estimated as the turnover time. This is approximately the process volume divided by the volumetric injection flow and the agitator pumping rate [6]. The largest dead time is usually the sum of transportation delay and the mixing time delay. This process dead time can be approximated by Equation 4-9.

$$\theta_p \cong \frac{V_i}{F_i} + \frac{V_b}{F_i + F_a} * K_z \qquad (4\text{-}9)$$

where:

F_i = volumetric injection flow (m^3/sec)

F_a = volumetric agitator pumping rate (m^3/sec)

K_z = geometric factor (1.0 for baffled vertical well mixed tank)

θ_p = process dead time from injection and mixing (sec)

V_i = injection (piping and dip tube) volume (m^3)

V_b = volume (m^3)

The largest source of loop dead time are heat transfer surfaces and sensor time lags, and mixing and transportation time delays.

For a change to be readily distinguished from noise, the excursion in the process variable must exceed the resolution limit and noise band of the measurement. This dead time is on average about one half of the resolution limit or noise band, whichever is significantly larger, divided by the rate of change of the process variable. The resolution limit for an analog-to-digital (A/D) converter with one signed bit is about 0.05 percent and 0.003 percent of span, for a 12-bit and 16-bit A/D, respectively. If the process variable is changing at 0.0001 percent per second (which is not uncommon for the composition response of the liquid phase), then the dead time can be estimated as 250 seconds and 15 seconds for a 12-bit and 16-bit A/D, respectively. The actual dead time can be almost zero, or twice as large as these numbers, because it depends on whether the measurement's starting point is at the opposite or near end of the resolution limit before the change.

The dead time from the resolution limit of the measurement is variable and significant for slowly changing process variables and 12-bit A/D converters.

In practice, the total loop dead time should be estimated based on the time between an appreciable change in the controller output or set point and the first noticeable change in the measurement. If there are no set point changes with the controller in automatic or output changes with the controller in manual, then you can roughly estimate the minimum dead time as the time delay from the start of the flow set by the batch sequence to the start of the measurement's response.

The time between the start of flow and the start of the measurement's response provides an estimate of the minimum loop dead time.

The observed dead time increases as the rate of change in controller output decreases. This is principally because of the increase in dead time from a slower ramp rate through the final element resolution limit and through the measurement resolution limit and noise band. The sensor may also respond more slowly because of the smaller gradient at the tip, particularly if there is any coating. A small set point change has the same effect since it corresponds to a small change in the controller output. The observed dead time for the gradual changes in load is much larger than the dead time observed for set point changes. This is because the change in controller output over the loop's response time is miniscule. The control error over the loop's response time, following from Equation 4-4, is incredibly small. The result is even significant fractional improvement in performances has a minimal effect, as Equation 4-2 suggests. This explains why even with far less than optimal tuning, loops still draw a straight line.

The steepness of the slope, rather than the change in slope of the batch profile the PID controller has to do. In other words, the magnitude of the controller output's rate of change (i.e., the size of shift in controller output within a batch phase) is a better indicator of load than a change in the controller's output rate of change. The tuning settings may change at different points in the batch, but the job for the PID is greatest at the controller output's maximum rate of change.

Limit Cycles

A resolution limit anywhere in the control loop causes a limit cycle. The period (T_0) of the limit cycle depends on the loop gains and the controller tuning settings, according to Equation 4-10, developed by Hägglund [8]. The limit cycle may be hidden by data compression, attenuated by signal filtering, and disguised by noise:

$$T_0 = 4 * T_i * [1/(K_{mv} * K_{pv} * K_{cv} * K_c) - 1] \tag{4-10}$$

The amplitude of the limit cycle depends on the final element and process gains, the limit cycle period, and the mixing time constant (τ_m), following from Equation 4-11 [9].

$$A_0 = (S_r * K_{mv} * K_{pv}) * [T_0/(2 * \pi * \tau_m)] \tag{4-11}$$

where:

A_o	=	limit cycle amplitude (%)
S_r	=	signal resolution of final element (%)
K_c	=	controller gain (dimensionless)
K_{cv}	=	controlled variable gain per measurement span (100%/Span PV e.u.)
K_{mv}	=	manipulated variable gain (Flow e.u./%)
K_{pv}	=	process variable gain (PV e.u./Flow e.u.)
T_i	=	controller integral time setting (seconds/repeat)
T_o	=	limit cycle period (seconds)
τ_m	=	mixing time constant for back mixed volume (seconds)

4-3. Self-Regulating Processes

The process response is fast and self-regulating for the secondary loops in column, reactor, and vessel cascade control systems. Flow, speed, and coolant temperature reach relatively quickly a new steady state in response to a manual change in the secondary controller's output.

Secondary loops generally have a fast self-regulating process response.

Figure 4-2 shows the open loop self-regulating response of the process to a manual change in the controller output. It is termed open loop because the controller is in manual, and it is termed self-regulating because the process decelerates to a new steady state. The process response in Figure 4-2 can be described by a steady-state process gain, a loop dead time, and a single time constant. This approximation is called a "first-order" model.

The loop dead time (θ_o) is the time from the start of the change in the controller output to the time at which an excursion of the process variable goes beyond its noise band. The time from the start of the change of the process variable to about 63 percent of its final value is the open loop time constant (τ_o). The final change in process variable (in percent) divided by the change in manual controller output (in percent) is the steady-state process gain (K_p).

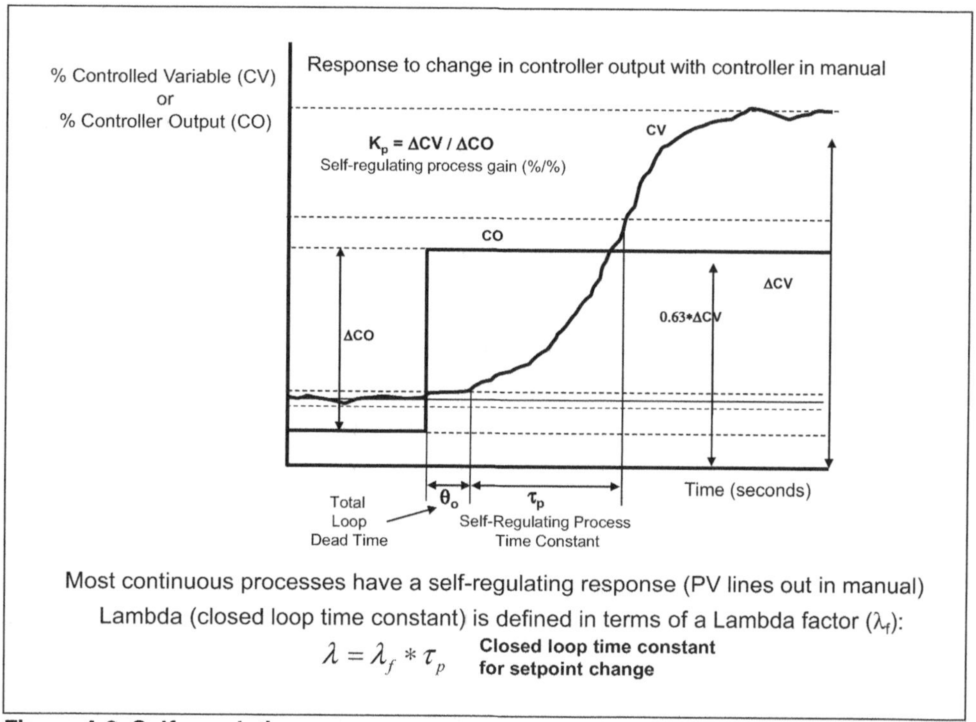

Figure 4-2. Self-regulating process response

Loop dead time delays the controller's ability to recognize and deal with change. If there was no noise, perfect control would be possible if the dead time were zero, in that the controller could immediately see and compensate for a change. The controller would be stable for even the most aggressive tuning settings.

If the loop dead time were zero, perfect control would be theoretically possible.

For the control of liquid flow and agitation by manipulating a variable speed drive (VSD) with a negligible dead band, the process dead time is small, and most of the loop dead time is set by the module execution time and resolution of the VSD. However, the improvement in control of the secondary loop from PI module execution times of less than one second is overshadowed by other considerations, such as proper tuning of the PI controller. The module execution times of primary controllers should be larger than one second to maximize the signal-to-noise ratio. This is because the real change over one second is within the noise band. Simply stated, the true change of the process variable within the module execution time should be five times larger than the noise. Larger module execution times are also desirable because they reduce controller loading, but the execution time must be less than twice the oscillation period to avoid severe aliasing. The primary module execution time should also be less than one-fifth the process dead time and time constant, whichever is smallest. The input and output cards and measurement devices are normally set up for over sampling in order to provide five scans for the fastest module execution time.

The module execution times of primary loops should be longer than one second to maximize the signal-to-noise ratio and minimize the loading.

A potentially large source of loop dead time for controlling coolant temperature is the thermal time lags for heat transfer between media and piping transportation delays. These lags and delays depend heavily on the coolant system design. Another significant source of dead time is the thermowell lag time, which varies from six to sixty seconds depending on the fluid velocity, fit of the sensing element within the well, and degree of coating [3]. For slowly changing process variables, the dead time that results from the final element's resolution is significant. However, it can be minimized by using sliding stem valves with digital positioners or a special high-resolution VSD [4] [5]. The resolution of a standard variable frequency drive is about 0.1 Hz in 60 Hz, which is about 0.17 percent. The resolution of rotary valves varies from about 0.25 to 5.0 percent depending on the degree of shaft windup and the type of positioner, packing, seating surfaces, and feedback mechanism. The resolution of sliding stem valves that have low friction packing and digital positioners typically ranges from 0.1 percent to 0.5 percent [4].

The heat transfer and transportation delays set during the design of coolant systems are the largest sources of dead time for coolant temperature control.

The open loop time constant can be approximated as the largest time constant in the control loop wherever it is located. The open loop time constant slows down the process variable's excursion rate and gives the controller a chance to catch up with a load change. If a large time constant in the process downstream of the point where disturbances enter is larger than the loop dead time it will improve control by slowing down these disturbances and by effectively filtering out process noise. Stated another way, a small dead-time-to-time-constant ratio improves the potential for good control within the loop. What is achieved in practice depends on the tuning settings, as seen in Equation 4-2. Often, a very large time constant is problematic because open loop tests that wait until a steady state is reached take too long. The time to reach 98 percent of the final response is approximately four time constants plus the dead time, as shown in Equation 4-12. Often the reason a loop is difficult is that the loop is too slow even if the slowness is in the process time constant, which offers the potential for better control.

$$T_{98} = 4 * \tau_o + \theta_o \tag{4-12}$$

where:

T_{98} = time to 98% of final response (sec)

θ_o = total loop dead time (sec)

τ_o = open loop time constant (sec)

Although a large process time constant might offer improved control of a properly tuned single loop controller, it can cause problems if it is in the secondary loop of a cascade control system. If the time constant in a secondary loop is not significantly less than the time constant in the primary loop, a fundamental rule is potentially violated: the secondary loop must be much faster than the primary loop. Some negative consequences can be avoided by using tuning methods that make the secondary loop faster. The section on cascade control in Chapter 3 addresses this concern in detail and provides a solution.

Although a large process time constant can improve the potential for better loop performance, it creates tuning and cascade control issues.

A large time constant caused by a sensor lag time, damping adjustment in the transmitter, or filter on the process variable in the control system also smooths out noise. However, it attenuates the control system's view of the true changes in the process. A large time constant in the measurement path creates an illusion of better control [3] [4] [5]. It also delays the control loop's ability to recognize a load upset.

Measurement time constants (e.g., filter times) give an attenuated version of the real process variability and an illusion of better control.

A controller that has too high a gain or rate action inflicts disturbances on itself by increasing the noise in the controller output. You can determine whether noise in the controller output is insufficiently smoothed out by the process time constant by determining whether the noise with the controller in automatic is greater than the noise when the controller is in manual. The solution to reduce noise introduced by automatic action is to first eliminate excessive gain or rate action and then if necessary increase signal filtering [5].

Noise introduced by automatic action should be first reduced by better tuning and then if necessary by adding a minimal filter on the process variable.

In reality, there is more than one time constant, which gives rise to a bend in the initial response, as seen in Figure 4-2. In the bend, the process variable accelerates. Subsequently, the response reaches an inflection point at which point it decelerates. If there was a single time constant, the response immediately after the dead time would be fastest and would form a right angle. For a "first-order" approximation, the small time constants effectively become dead time. This is determined by the intersection of the process variable's initial value by a line that is tangent to the inflection point, as shown in Figure 4-2.

In real applications, there are many small time constants that can be approximated as additional dead time.

Adding rate action can compensate for these small time constants if the measurement noise is small and a large process time constant exists to smooth out the reaction to noise in the controller output. Flow and speed loops have too much noise and no appreciable process time constant. Composition loops may have too much sensor noise. The best candidates for rate action are generally temperature loops that have negligible A/D noise. When one-way cooling is used (no heating via a tempered water system) and there is no appreciable heat of reaction, then it is essential that derivative action be used to prevent overshoot of the set point.

The best candidates for rate action are generally temperature loops with minimal noise, in which the rate time is set to compensate for small time constants.

The process gain is more linear for flow and speed loops that have a VSD since these loops do not have the flow characteristic of a control valve. The process gain for gear pumps is relatively constant and does not vary much with operating point or process conditions. The steady-state process gain for a temperature loop that throttles coolant flow is inversely proportional to coolant flow. The equal percentage flow characteristic of a control valve can largely cancel out this process gain nonlinearity. If a VSD, linear valve, or secondary flow loop is used, then there is no inherent compensation. At low loads, the process gain increases dramatically. Also, at low flow, the transportation delay and heat transfer lag increase. This combination of a higher process gain and larger loop dead time can cause a limit cycle in a temperature control system. Whether it is noticeable or not depends on data compression, mixing, and tuning.

If the flow loop manipulates a VSD without velocity limits instead of a control valve as the final element, then the time constant is dramatically faster. The loop's speed of response is mostly set by the module execution time, transmitter damping, and signal filters.

4-4. Integrating Processes

The process response is slow and integrating for primary batch loops. The temperature control of batch operations that have slow reaction rates resembles an integrating response because no discharge flow occurs for continuous control. Also, in batch operations, there is no steady state as evidenced by the batch profiles. After the dead time, the process slowly ramps in response to a manual change in the primary controller output, as shown in Figure 4-3. There is no steady state and consequently no steady-state process gain for these processes. The process gain is an integrating gain that is the change in ramp rate in %/sec per percent change in the controller output. The response is modeled as a loop dead time and an integrating process gain. Note also that the initial response is not flat for manual operation as it was for a self-regulating process. This is because a small unbalance (difference between load and controller output) causes the process variable to ramp. Thus, the process gain must be calculated from the change in ramp rate.

Primary batch loops typically have a slow integrating type of process response.

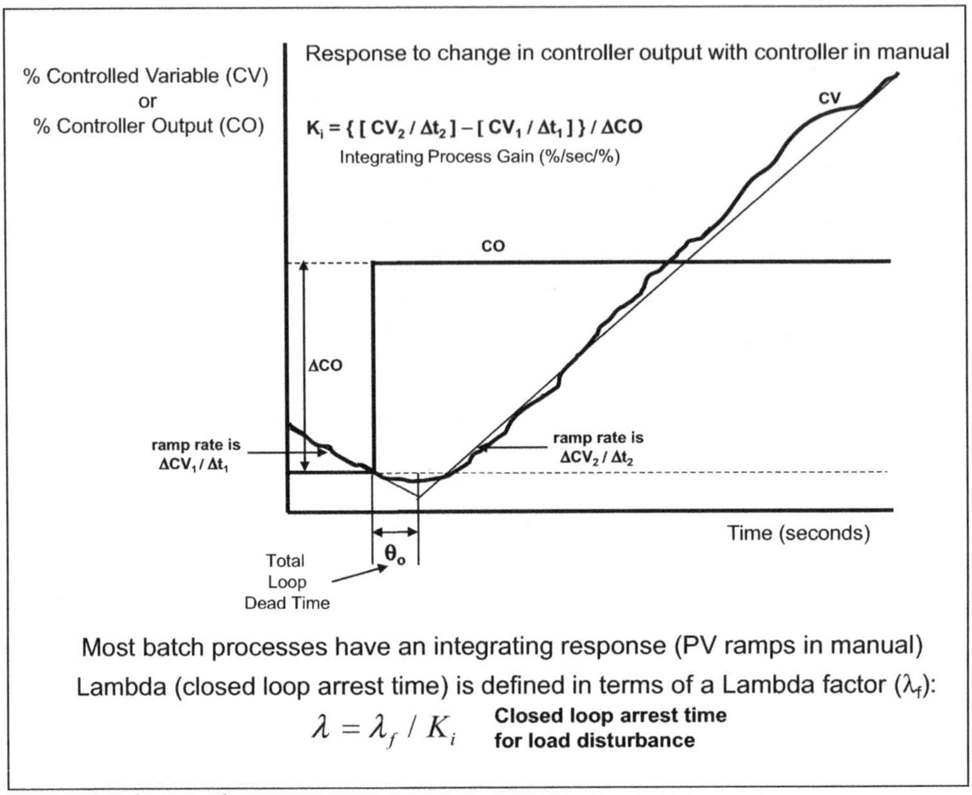

Figure 4-3. Integrating process response

An integrating process does not line out when the controller is in manual.

If the process output's ramp rate (slope) is used for the controlled variable where the process gain is the change in the process output's ramp rate (slope) for a change in the process input, then the result can be modeled as a self-regulating response. When the ramp rate is constant for a given change in the controller output, there is effectively a steady state. However, this ramp rate changes as the batch progresses.

An integrating process can be modeled as a self-regulating process with an interim steady state by computing the change of the process output's ramp rate in response to a change in a process input.

Integrating processes can be identified much faster because there is no steady state. The person or software must just wait until the ramp rate is recognizable, which is usually possible after four dead times.

Integrating processes can be identified faster than self-regulating processes.

Disturbances also change the ramp rate. The dead time and process gain change in relation to operating point and process conditions. The dead time may also change as a function of the direction of the change. For these and other reasons, tests should be repeated at least twice in both directions at different conditions.

Tests to identify process dynamics should be repeated at least twice in both directions at different operating points, process conditions, and batch times.

Though a batch temperature loop may theoretically have a self-regulating response, the time required to reach steady state is so long and the self-regulation so weak because of a lack of continuous discharge flow that it is best to model and tune the response as an integrating process. This can be visualized by noting that the first portion of the response in Figure 4-2 after the dead time is similar to the response in Figure 4-3. Processes that do not have a true integrating response but ramp in the time frame of interest are called "pseudo" or "near" integrating processes [3] [5]. Equation 4-13 shows the conversion of the steady-state open loop gain (often just referred to as process gain) and the open loop time constant of a self-regulating process into the equivalent integrating process gain of a near integrating process [3] [5]. Modeling a slow self-regulating loop as a near integrating process can save considerable test time. Chapter 3, on basic feedback control, discusses how to apply the tuning rules for integrators to these processes.

Batch operations tend to have an integrating type of concentration and temperature response because there is no continuous discharge flow or steady state.

$$K_i = K_o \ / \ \tau_o \tag{4-13}$$

$$K_o = K_{mv} * K_{pv} * K_{cv} \tag{4-14}$$

where:

K_i = "pseudo" or "near" integrating process gain (%/sec/%)

K_o = steady-state open loop gain (also known as process gain) (%/%)

K_{cv} = controlled variable gain per measurement span (100%/Span PV e.u.)

K_{mv} = manipulated variable gain per pump or valve (Flow e.u./100%)

K_{pv} = process variable gain (PV e.u./Flow e.u.)

τ_o = open loop time constant (also known as process time constant) (sec)

As was the case for the self-regulating response, the initial bend in the response is associated with small time constants. If the process could be modeled by a loop dead time and an integrating process gain, then the process variable would abruptly reverse immediately after the dead time. In real applications, the transition is smoothed by small time constants. A rate time can be set to compensate for these time constants if the measurement noise is not excessive and the integrating process gain is low, which, according to Equation 4-13, is equivalent to a large process time constant.

An exothermic reactor can have a runaway temperature response at high temperatures where the temperature excursion accelerates as shown in Figure 4-4. The exponential increase in the heat of reaction with temperature causes a positive feedback time constant. Open loop tests are not held long enough to see the acceleration due to safety concerns.

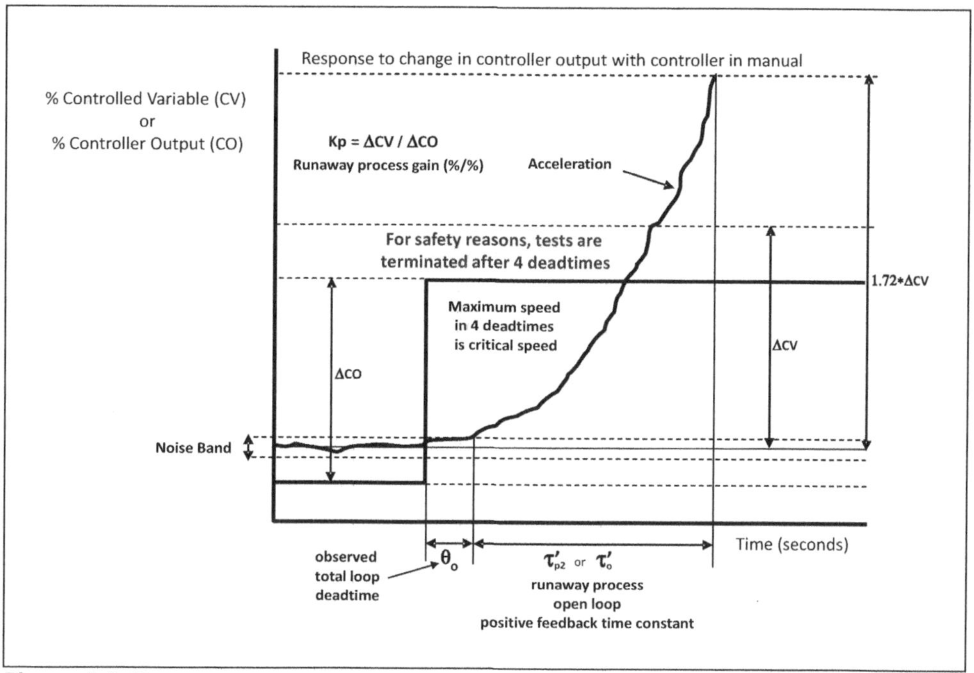

Figure 4-4. Runaway response

Exercises

4-1. What is the largest source of dead time in a small control valve?

4-2. If you double the controller gain, what happens to the integrated error for a disturbance provided the controller is not oscillating?

4-3. What are the major problematic measurement sources of loop dead time?

4-4. If you double the reset time, what happens to limit cycle amplitude after filtering by a well mixed volume?

4-5. What translation of the control variable allows a integrating process be modeling as a self-regulating process for batch profile control?

References

1. McMillan, Gregory K., *Tuning and Control Loop Performance*. 3d ed. ISA, 1992.

2. Blevins, Terrence L., McMillan, Gregory K., Wojsznis, Willy K., and Brown, Michael W., *Advanced Control Unleashed: Plant Performance Management for Optimum Benefits*. ISA, 2003.

3. McMillan, Gregory K., *Good Tuning: A Pocket Guide. 2d ed.* ISA, 2005.

4. McMillan, Gregory K., Sowell, Mark, and Wojsznis, Peter W., "The Next Generation – Adaptive Control Takes a Leap Forward." *Chemical Processing*, September 2004.

5. Hägglund, T., "A Control Loop Performance Monitor." *Control Engineering Practice* 3, no. 11 (1995): 1543-51.

6. McMillan, Gregory K., "What Is Your Control Valve Telling You?" *Control Design*, May 2004, 43-48.

5
Exchangers

This chapter describes control strategies and operational considerations for shell and tube heat exchangers.

Learning Objectives

A. **Understand how some exchanger designs make temperature control nearly impossible.**

B. **Know how to compensate or adapt for changes in the process gain, dead time, and time constant.**

C. **Recognize the limitations of some control schemes.**

D. **Be able to construct the best control strategy.**

5-1. Process and Equipment Design Considerations

If either of the ratios X_1 and X_2, as defined by Equations 5-1, 5-2, 5-4 and 5-5 and illustrated by associated Figures 5-1 and 5-2, are greater than three, the change in the controlled temperature with manipulated flow is often too small for good control (for the counterflow liquid – liquid heat exchanger depicted in Figure 5-3) [1]. This loss of process sensitivity will show up as prolonged deviations of the measurement from set point for increases in head load. Very high controller gains and feedforward control can help, but the loop often runs out of valve (i.e., the control valve is wide open and controller output is at its output limit). Differences in exchanger design and process conditions may shift this point, but the problem remains the same. The system should be designed to eliminate high flow operation to avoid insufficient process sensitivity and high energy usage from larger coolant pressure drops and flow.

The ordinate Y, per Equations 5-3 and 5-6, is the temperature drop, or rise from inlet to outlet on the controlled side, normalized by division by the maximum possible temperature change, which is the difference between the hot and cold inlet temperatures [1]. The slope of Figures 5-1 and 5-2 is representative of the process gain. Thus, the process gain increases as the manipulated flow decreases. The slope gets exceptionally steep as the coolant flow approaches zero, and the transportation delay gets large. The exchanger and control system should be designed to eliminate low flow operation to avoid excessive process sensitivity and the dramatic increase in fouling of the heat transfer surfaces that occurs at low velocities.

Flows much lower than design causes a high process gain and a high process deadtime that can lead to temperature oscillations and possible instability.

Flows much higher than design causes a low process gain (poor sensitivity) that can lead to wandering of the temperature and possible loss of temperature control.

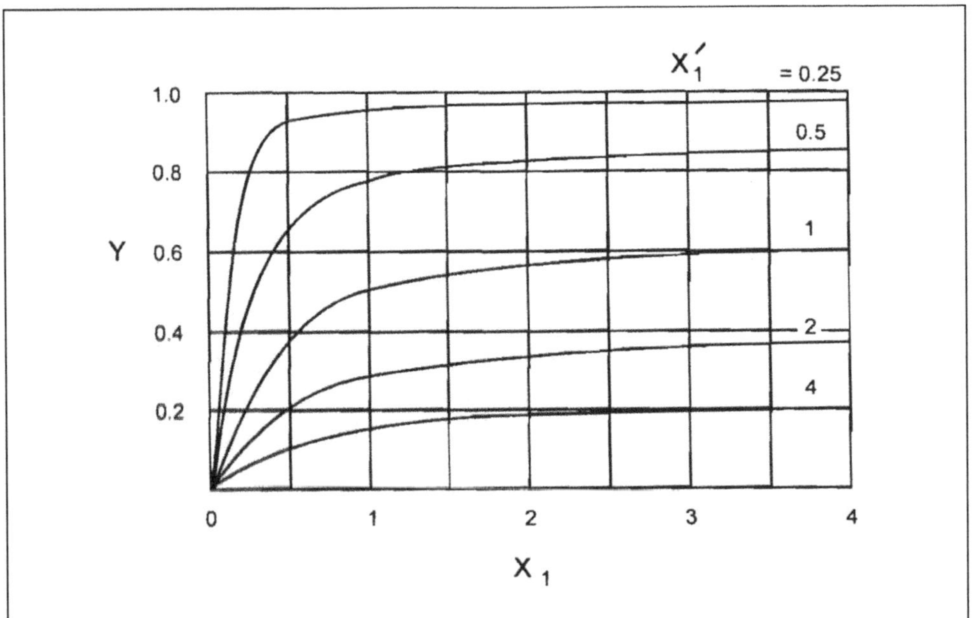

Figure 5-1. If the heat transfer area is too small or large, the process gain will be too small or large, respectively [1]

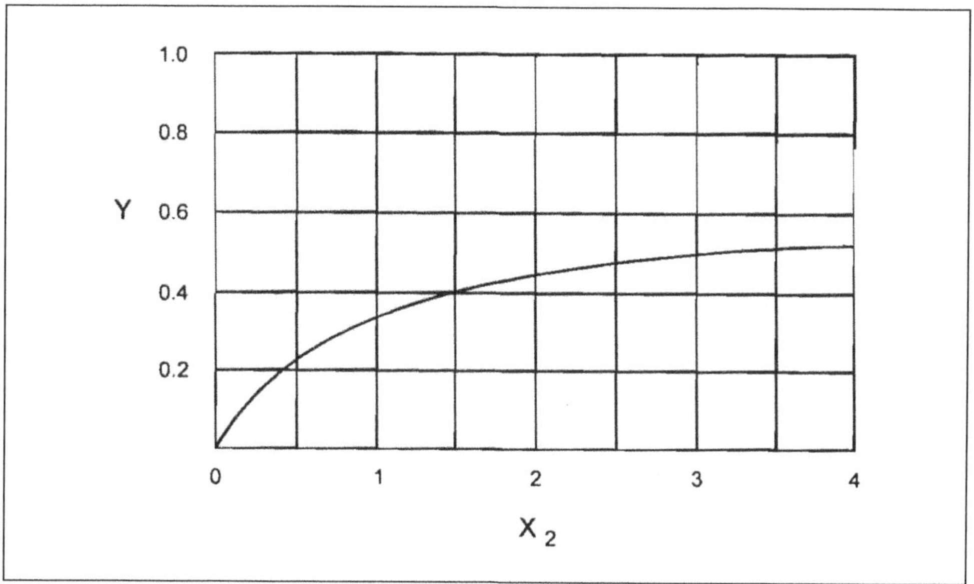

Figure 5-2. If the uncontrolled flow is too small or large, the process gain will be too large or small, respectively [1]

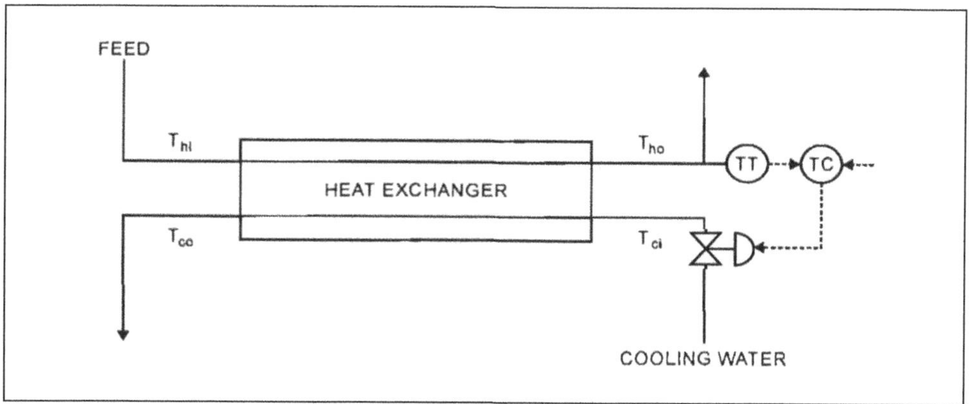

Figure 5-3. In counterflow liquid – liquid exchangers, the temperature difference at each end of the exchanger is maximized by pairing the cold inlet with the hot outlet and the cold outlet with the hot inlet

Thus, the mechanical process design that sets the heat transfer areas and geometry can make the control easy or difficult. While advanced control methods can treat the symptoms, the better solution is to eliminate the root cause of the problem by early involvement of the process control engineer in the exchanger design. Note that X'_1 is the X_1 for the side not manipulated. Thus, if Equation 5-1 is used for X_1, then Equation 5-4 is used for X'_1, and vice versa.

For control and manipulation of cold side:

$$X_1 = (F_c * C_c)/(U * A) \tag{5-1}$$

$$X_2 = F_c/F_h \tag{5-2}$$

$$Y = (T_{ci} - T_{co})/(T_{hi} - T_{ci}) \tag{5-3}$$

For control and manipulation of hot side:

$$X_1 = (F_h * C_h)/(U * A) \tag{5-4}$$

$$X_2 = F_h/F_c \tag{5-5}$$

$$Y = (T_{ho} - T_{hi})/(T_{hi} - T_{ci}) \tag{5-6}$$

where:

A = heat transfer area (ft^2)

C_c = cold side fluid heat capacity (Btu/°F*lb)

C_h = hot side fluid heat capacity (Btu/°F*lb)

F_c = cold side mass flow (lb/hr)

F_h = hot side mass flow (lb/hr)

T_{ci} = cold side inlet temperature (°F)

T_{co} = cold side outlet temperature (°F)

T_{hi} = hot side inlet temperature (°F)

T_{ho} = hot side outlet temperature (°F)

U = overall heat transfer coefficient (Btu/hr*ft^2)

X_1 = abscissa for process gain plot 1

X_2 = abscissa for process gain plot 2

Y = ordinate for process gain plots

For condensing streams (e.g., steam or column overheads), a smaller control valve can be used by throttling the condensate drain as shown in Figure 5-4. Since the heat transfer coefficient for condensing vapor is much higher than for a low velocity liquid, an increase in liquid level that covers the heat transfer area will decrease the heat transfer and raise the controlled temperature. However, this puts an integrator (i.e., level) in a con-

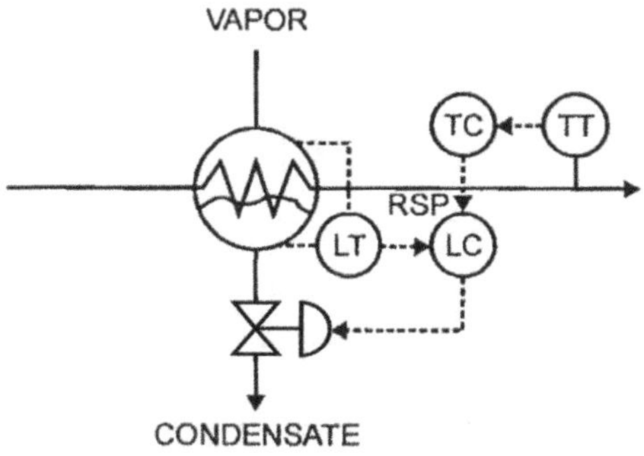

Figure 5-4. A secondary level loop can improve condenser temperature control

trol loop that has significant loop dead time of about one minute or more due to heat transfer lags and the measurement lag. For small diameter condensers, the loop could always become unstable if the level change during the loop dead time is a large portion of the level span. Furthermore, the speed of a level decrease may be faster than a level increase because the condensate removal rate can be much larger than the condensation rate. The result is poor control.

The addition of a secondary level loop makes the level response for temperature control self-regulating. If the level controller gain is maximized, the improvement in control is significant.

5-2. Disturbances and Difficulties

The control of a liquid – liquid exchanger outlet temperature by manipulation of the bypass of hot flow around the exchanger per Figure 5-5 keeps the total coolant flow fixed, but the flow through the exchanger is still variable. While three-way valves are often used, the author favors separate valves, as shown, for better throttling flexibility, performance, and diagnostics. The operating point should still be kept below a ratio of three, because the heat transfer, as indicated by the temperature difference *Y*, becomes nearly constant at higher ratios. If the heat transfer is constant, the temperature of the combined streams at the outlet also remains constant. Thus, the same nonlinearity problem still exists. The optimum ratio lies between a ratio of one-half to two to keep the coolant flow high enough to reduce fouling but not so high as to reduce loop sensitivity.

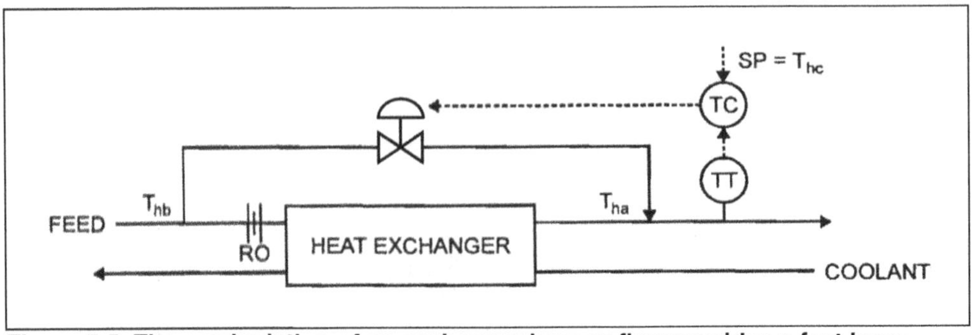

Figure 5-5. The manipulation of an exchanger bypass flow provides a fast loop; however, the measurement lag is the largest time constant, and the process gain nonlinearity still exists [1]

Bypass heat exchanger control is fast, but the flow through the exchanger must be kept high enough to reduce fouling.

The fraction *K* of the residence time that becomes dead time typically varies from about 0.2 to 0.8, depending upon the exchanger geometry and flow rate. The remaining portion of the residence time is the time constant. The dead time to time constant ratio increases as the flow decreases or the length-to-diameter (L/D) ratio increases due to a decrease in back mixing. Exchangers with long narrow tubes or low flow will have a greater portion of the residence time show up as dead time than in designs with shorter larger tubes and high flow.

5-3. Effect of Loop Performance

Exchangers are often a subsystem of a unit operation. Poor exchanger control adversely affects the temperature control in processing equipment that is important for quality control. For example, temperature oscillations from exchangers in reactor or crystallizer recirculation lines, or distillation column feed lines, can cause offspec product. The rapid severe oscillations of limit cycles from low coolant flow excursions are particularly disruptive. Interaction between the exchanger and process equipment temperature loops can result in continuous cycling of all the temperature loops. For this reason, exchanger bypass or coolant makeup are preferred over coolant throttling as methods of liquid – liquid exchanger control For throttling steam to exchangers, jackets, or coils, a pressure controller or remotely set pressure regulator is recommended as the inner (secondary) loop of a cascade control system [3].

5-4 Controller Tuning

The increase in loop period, the decrease in process time constant, and the increase in process gain require a dramatically lower controller gain for low flow. For an X ratio less than one, a controller gain tuned for quarter amplitude response may have to be decreased by a factor of four to ten for a reduction of 50% in flow. The factor of four corresponds to a more dead time dominant system (e.g., tube side throttled) where the controller gain becomes more dependent upon the open-loop gain than on the dynamics [2]. Limit cycles develop for excursions below a ratio of one. Some users have found out experimentally that elevation of the controller low output limit helps. The use of an equal percentage trim can cut the gain change in half, because the valve gain becomes proportional to flow. The improvement is greater for a more constant pressure drop across the valve stays, so that the installed characteristic is closer to the inherent trim characteristic. The benefit of an equal percentage characteristic can be achieved more accurately by application of the signal divider, detailed in Equations 5-7 and 5-8, to the temperature controller output.

$$Y = X/[Z + (1 - Z) * X] \tag{5-7}$$

$$Z = (1/R)^{0.5} \tag{5-8}$$

where:

X = signal divider input (%)

Y = signal divider output (%)

Z = rangeability factor (dimensionless)

The proper solution is adaption of the temperature controller gain and reset settings. Equations 5-9, 5-10, and 5-11 provide approximate rules of thumb for adaption of PID controllers for control of liquid – liquid exchangers with an X ratio of less than one and manipulation of either a valve with a linear installed characteristic or a flow controller set point. The derivative setting (i.e., rate time) doesn't need adaption unless the residence time is much larger than two minutes or a bare element is used, because the setting depends mostly upon the secondary measurement time constant (e.g., thermowell lag).

For tube side throttling:

$$K_c' = (F'/F)^2 * K_c \tag{5-9}$$

For shell side throttling:

$$K_c' = (F'/F)^3 * K_c \tag{5-10}$$

For tube or shell side throttling:

$$T_i' = (F'/F)^1 * T_i \tag{5-11}$$

where:

F = normal throttled flow (gpm)

F' = actual throttled flow (gpm)

K_c = normal controller gain (dimensionless)

K_c' = actual controller gain (dimensionless)

T_i = normal controller integral time (minutes/repeat)

T_i' = actual controller integral time (minutes/repeat)

Adaptive control where tuning settings are scheduled as a function of manipulated flow can compensate for the increase in process gain and dead time as the throttled flow decreases.

Most industrial exchangers have a total loop dead time of thirty seconds to two minutes and thus a period of one to eight minutes and a reset setting of 1.0 to 0.1 repeats per minute. The loop dead time is the sum of process dead time, the controller scan time, the dead time from valve dead band, and the portion of the thermowell time constant that becomes dead time. The process dead time is negligible for manipulation of an exchanger bypass (Figure 5-5). For direct throttling of exchanger low, the process dead time is the portion of the residence time and thermal lag across the metal heat transfer surfaces converted to dead time. The metal thermal lag is the product of the metal mass and heat capacity divided by the product of the overall heat transfer coefficient and area. This time constant is usually small for most exchangers (e.g., 0.1 minute). The sum of the sensor and thermowell measurement time constants is larger (e.g., 0.2 to 0.8

minute). Since the derivative time is normally set equal to the sum of these secondary time constants in the loop, temperature controllers for exchangers should have a nominal derivative or rate setting of 0.3 to 0.9 minute that largely depends upon the thermowell material and air gap. For jackets and coils in reactors, the metal lag and other secondary lags are larger, which translates to larger derivative settings for the reactor temperature controllers.

The process gain is inversely proportional to the manipulated exchanger flow. For large shells, the interaction with the other side can be ignored, and the process gain simplifies to the product of Equations F-6i and F-9a in Appendix F, which is used for reactors, fermentors, and vessels. Otherwise, the energy balance on both sides and the effect of the log mean temperature difference should be included. Equations 5-12 and 5-13 provide a steady solution of the outlet temperature, given the flows and other temperatures. It can be solved for the outlet temperature of choice (i.e., controlled variable) and then differentiated with respect to the manipulated flow to give the process gain. An easier and more visual solution is to plot the steady state solution versus the ratio of manipulated to feed flow in an Excel file. The process gain is the slope of this plot divided by the feed flow. Since the conduction heat transfer resistance is negligible and the heat transfer coefficient changes with 0.8 power of flow, the variation in the heat transfer coefficient can be approximated by Equations 5-14 through 5-16.

The controlled temperature should be plotted versus the ratio of manipulated flow to feed flow so that the changes in slope and, hence, the nonlinearity of the process gain can be identified.

$$T_{ho} = T_{ci} + e^{m} * (T_{hi} - T_{co}) \tag{5-12}$$

$$m = (U * A)/(F_c * C_c) - (U * A)/(F_h * C_h) \tag{5-13}$$

$$U = (h_c * h_h)/(h_h + h_c) \tag{5-14}$$

$$h_c = F_c^{0.8} * f_c \tag{5-15}$$

$$h_h = F_h^{0.8} * f_h \tag{5-16}$$

where:

A = heat transfer area (ft^2)

C_c = cold side fluid heat capacity (Btu/°F*lb)

C_h = hot side fluid heat capacity (Btu/°F*lb)

F_c = cold side mass flow (lb/hr)

F_h = hot side mass flow (lb/hr)

f_c = cold side mass heat transfer factor (Btu/lb*ft^2)

f_h = hot side mass heat transfer factor (Btu/lb*ft^2)

h_c = cold side heat transfer coefficient (Btu/hr*ft^2)

h_h = hot side heat transfer coefficient (Btu/hr*ft^2)

m = base e exponent (dimensionless)

T_{ci} = cold side inlet temperature (°F)

T_{co} = cold side outlet temperature (°F)

T_{hi} = hot side inlet temperature (°F)

T_{ho} = hot side outlet temperature (°F)

U = overall heat transfer coefficient (Btu/hr*ft^2)

For exchangers where the inner loop is a steam pressure controller, the change in steam pressure corresponds to a change in heat transfer surface temperature. Thus, the process gain is the change in steam temperature divided by the change in steam pressure from steam tables, multiplied by the change in controlled temperature divided by the change in steam temperature at operating conditions.

5-5. Control Errors

The peak error is rather large for fast disturbances due to a time constant to dead time ratio that ranges from poor (e.g., 4.0) to lousy (e.g., 0.4). The larger ratios correspond to moderate flows and temperature control on the shell side where more backmixing occurs. The worse case is a small coolant flow and temperature control on the tube side with large transportation delays from low plug flow.

The integrated error is quite small despite large peak errors because the loop period is relatively small for temperature control. Thus, the use of exchanger temperature as an inner (i.e., slave or secondary) loop of a cascade control system for vessel temperature control is quite effective because the rapid oscillations are effectively filtered by the large process time constant of the liquid volume.

When peak error or initial transient must be minimized, feedforward control should be used. A rather simple energy balance that equates heat lost from the hot side to heat gain by the cold side yields a solution per Equations 5-17 (coolant flow) and 5-18 (steam flow) for the required final element flow for a given cold and hot inlet temperature, feed flow, and set point. Normal operating values are used for those inputs not measured that are relatively constant. For example, if the main upset is feed flow, a measurement of this flow is required, but assumed operating conditions can be used for the inlet temperatures that are not measure. the equations show temperature control of the hot side. For control of a cold side, only the subscripts need to be changed. It is critical that the controlled temperature be the set point rather than the measurement to avoid positive feedback.

It is critical that the temperature setpoint rather than the temperature measurement be used in the feedforward signal calculation to avoid positive feedback.

$$F_c = F_h * C_h * (T_{hi} - T_{ho}) / [C_c * (T_{ci} - T_{co})] \quad (5\text{-}17)$$

$$F_s = F_c * C_c * (T_{ci} - T_{co}) / H_s \quad (5\text{-}18)$$

where:

C_c = cold side fluid heat capacity (Btu/°F*lb)

C_h = hot side fluid heat capacity (Btu/°F*lb)

F_c = cold side mass flow (lb/hr)

F_h = hot side mass flow (lb/hr)

F_s = steam mass flow (lb/hr)

H_s = steam enthalpy (Btu/lb)

T_{ci} = cold side inlet temperature (°F)

T_{co} = cold side outlet temperature (°F)

T_{hi} = hot side inlet temperature (°F)

T_{ho} = hot side outlet temperature (°F)

The loop set point must be used for the controlled temperature!

The feedforward signal is added to the output of the feedback controller. A bias of 50% is also subtracted so that the temperature controller can make a negative correction as large as the positive correction to the feedforward signal. The manipulated flow is best achieved by means of a flow controller and a cascade of exchanger temperature to coolant or steam flow. If the temperature controller output goes directly to a control valve, signal characterization of the installed valve characteristic should be used to convert from desired flow to required valve position. The signal divider for compensation of process gain nonlinearities should be applied to the controller output before the summer, as shown in Figure 5-6. The divider corrects for the process gain being inversely proportional to flow. Of course, the coolant or steam valve should have a positioner. If the manipulated variable is a steam pressure controller or a regulator set point, the substitution of the steam temperature for the hot side temperatures of Equation 5-12 and the subsequent solution of the equation for steam temperature will yield a corresponding saturated steam pressure. Here the objective is to maintain the proper log mean temperature difference or driving force and let the pressure controller or regulator determine the actual steam flow needed. The feedforward loop may not be needed, because a fast inner steam pressure loop or regulator can correct for disturbances before they affect the temperature loop.

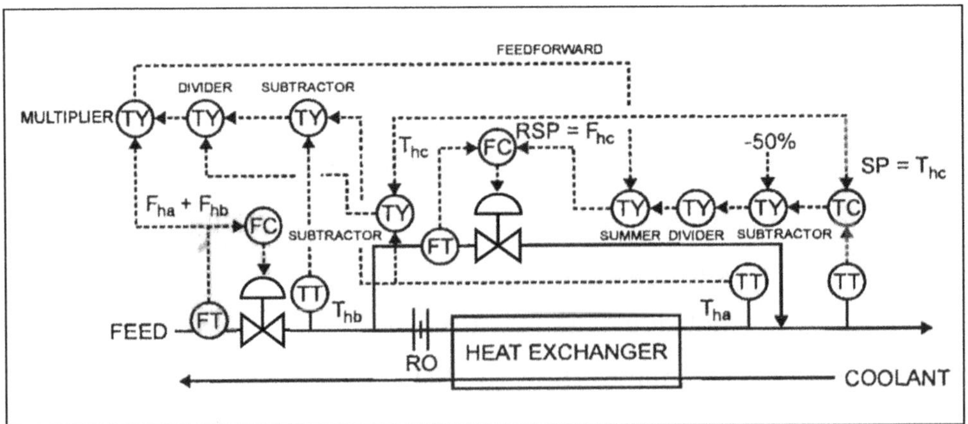

Figure 5-6. The temperature controller output provides a positive and negative bias correction to the feedforward signal that uses the temperature set point in an energy balance

If the exchanger outlet temperature can be adjusted, a valve position controller can be used to slowly change the temperature set point to optimize the coolant valve position, as shown in Figure 5-7. There may be an optimum position that minimizes the fouling of heat transfer surfaces by prevention of low throttle positions but also maintains the process gain and reduces utility usage by the prevention of high throttle positions.

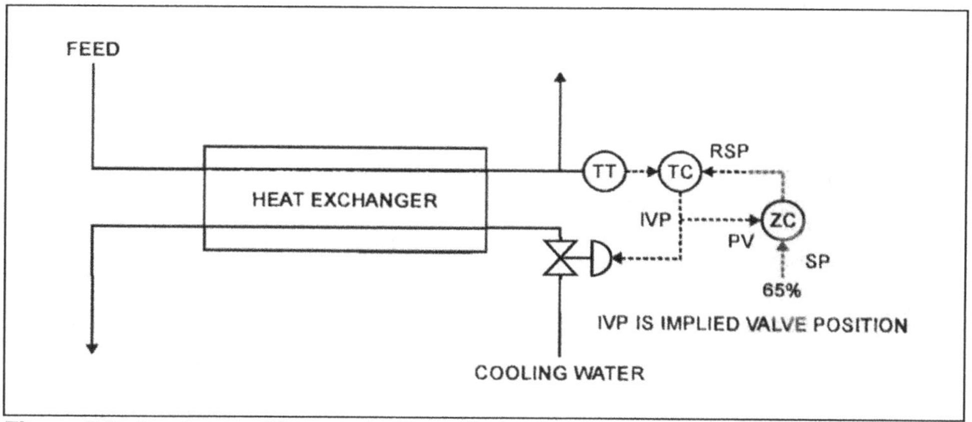

Figure 5-7. A valve position controller will slowly trim the temperature set point to minimize fouling of heat transfer surfaces at the best possible process gain and utility usage

For exchanger bypass control (see Figure 5-5), the blend of the hot and cold streams is achieved by a split of the total flow into an actual exchanger flow and a bypass flow. Since the heat capacities are approximately equal, the ratio of bypassed flow to total flow can be approximated as proportional to the ratio of the biased controller set point to the biased bypass temperature, per Equation 5-19. The bias is a subtraction of the actual exchanger outlet temperature. Note that the bypass temperature is also the inlet temperature to the exchanger.

$$R = F_{hb}/(F_{hb} + F_{ha}) = (T_{hc} - T_{ha})/(T_{hb} - T_{ha}) \quad (5\text{-}19)$$

where:

F_{ha} = exchanger's actual hot side mass flow (lb/hr)

F_{hb} = exchanger's bypass hot side mass flow (lb/hr)

R = ratio of manipulated flow to total flow

T_{ha} = exchanger's actual hot side temperature (°F)

T_{hb} = exchanger's bypass hot side temperature (°F)

T_{hc} = combined hot side temperature (°F)

5-6. Control Strategies

Table 5-1 summarizes some of the options. As usual, there is a tradeoff of cost for performance. It is the author's opinion that the increased cost of piping, control valves, or measurements is more than offset by either increased product quality or decreased energy consumption from fewer oscillations. Advanced control techniques such as nonlinearity compensation and feedforward can be used to treat the symptoms of a poor design or implementation. It is better to eliminate the root cause and provide an installation that is inherently good.

Table 5-1. Improved performance often requires higher initial costs but is justified by the reduced cost of goods (COGS)

Manipulated Variable	Installation Cost Rating and Reason	Loop Performance Rating and Reason
Coolant valve	Low: least piping and instrumentation	Lousy: large variable process and valve gain, dead time, flow upsets, also exchanger fouling
Coolant flow	Medium: flow measurement and controller	Poor: large variable process gain and dead time, also exchanger fouling
Bypass flow	Medium: piping and 3-way or dual control valves	Fair: large variable process gain
Coolant makeup (fixed flow)	High: extensive piping and circulation pump	Good: small process gain and dead time
Condensate valve	Cheap: small control valve	Lousy: slow fill and fast empty plus non self-regulation
Steam valve	Low: large control valve	Poor: variable valve gain, flow, steam, and load upsets
Steam flow	High: flow measurement and controller	Fair: steam and load upsets
Steam Pressure	Medium: pressure measurement and controller	Good: compensation of upsets and delta temperature control

Exercises

5-1. The process stream from a reactor is cooled using the heat exchanger below. Using Equations 5-1 and 5-5, determine if this stream can be controlled at the desired temperature.

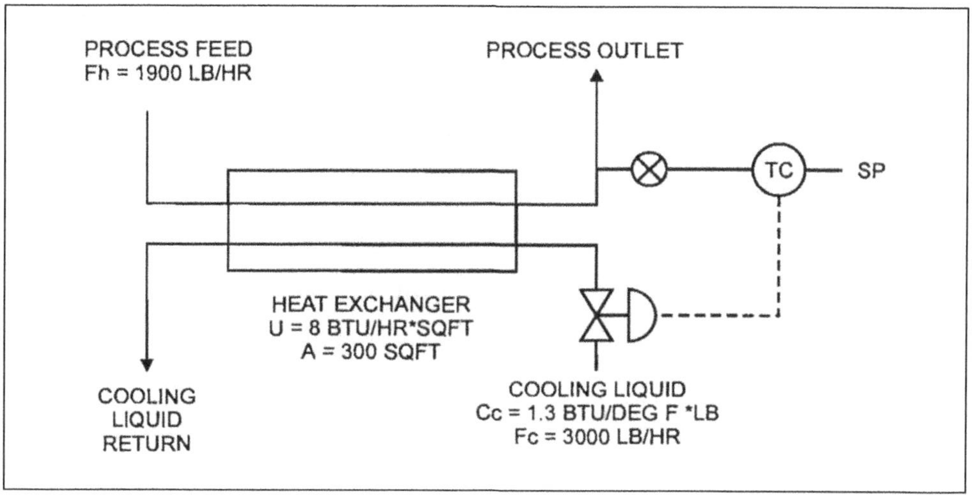

5-2. Which heat exchanger setup below offers the maximum heat transfer?

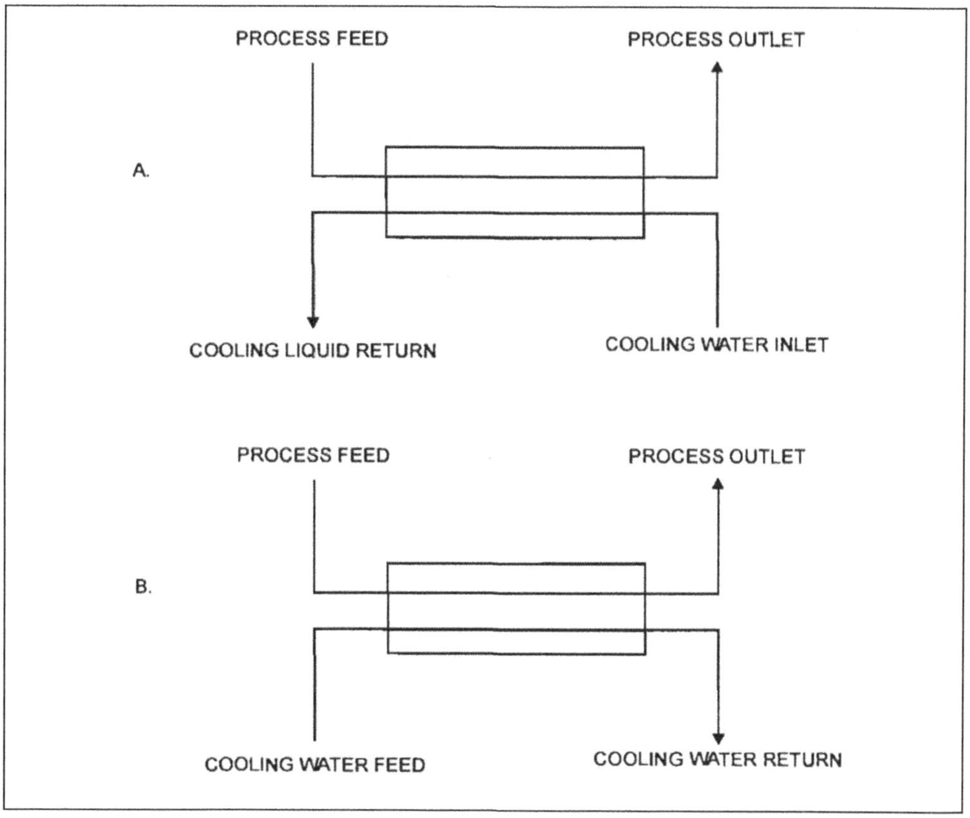

5-3. Why are heat exchangers so hard to control?

5-4. A liquid-liquid heat exchanger is designed for a normal coolant flow of 50 gpm. The tube side of the exchanger is being throttled, and the exchanger has an *X* ratio of 0.5 at nominal conditions. The temperature controller used on the exchanger is tuned with a gain set6ting of 0.3, reset value of 0.8 min./repeat and a derivative setting of 0.2 min. A process change requires that the exchanger operate with half the coolant flow. Using Equations 5-9 and 5-11, determine the new temperature controller settings.

5-5. A liquid-liquid heat exchanger is used to cool a process stream. A feedforward control scheme is used to control the temperature of the process stream. Using Equation 5-17, determine the coolant flow required to maintain the conditions below.

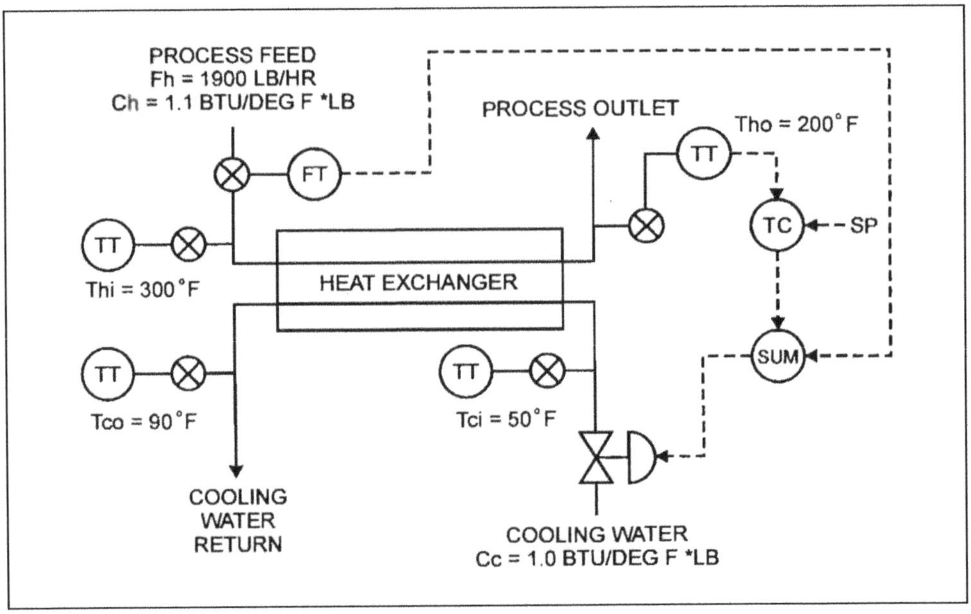

References

1. Shinskey, F. G., *Process Control Systems*, McGraw-Hill Publishing Company, third edition, 1988.

2. McMillan, G. K., *Tuning and Control Loop Performance*, ISA, third edition, 1994.

3. Spitzer, D. W., "Remotely Set Pressure Regulators: A User's Perspective," ISA Conference Paper 91-0464, October 1991.

6
Reactors

This chapter describes control strategies and tuning requirements for reactors.

Learning Objectives

A. Recognize when and how a poor coil, jacket, or thermowell design can make exothermic reactor control impossible.

B. Be able to correct the most common mistake in tuning temperature controllers for reactors.

C. Determine how to adapt the controller tuning for the different operating conditions of batch reactors.

D. Know how to set up the best cascade loop.

6-1. Process and Equipment Design Considerations

The process time constant is much larger than other time constants and dead times in the temperature control loop for a properly designed reactor. This assumes the following sources of dead time total less than ten percent of the process time constant:

1. The mixing turnover time (approximated as the volume divided by the sum of throughput, recirculation, and agitator pumping flow rates).
2. The heat transfer surface thermal time constant (approximated as the product of the surface mass and heat capacity divided by the product of the heat transfer coefficient and area).

3. The coolant transportation delay (approximated as the coolant volume divided by the coolant flow rate).

4. The thermowell-sensor time constant (see Section 2-3).

For this situation, the loop dead time is small, and excellent reactor temperature control is possible for constant coolant flow. Conversely, insufficient agitator pumping rate, a large heat transfer surface mass-to-area ratio, a low coolant flow, a large or glass-coated thermowell, or a loose sensor fit in a thermowell create enough equivalent dead time from secondary lags or pure dead time to cause poor control despite the best of tuning. A variable volume or a variable coolant flow will cause a process gain nonlinearity. The effect of changes in coolant flow are particularly disastrous due to the additional changes in coolant transport delay, as described in Section 5-5 for heat exchanger temperature control. To keep the coolant flow fixed, a coolant pump and recirculation system are needed. The coolant heat removal rate is changed by throttling a coolant makeup flow to change the coolant temperature. Normally, as shown in Figure 6-1, the output of the reactor temperature controller sets the remote set point of a coolant temperature controller. The cascade control of reactor to coolant temperature has excellent dynamics, in that the inner (secondary) loop is much faster than the outer (primary) loop. The inner loop corrects for disturbances in makeup coolant temperature or pressure and valve deadband before they affect the reactor temperature and provides a process gain that is simply the reactant temperature span divided by the coolant measurement span. The process gain is no longer dependent upon the temperature difference between the reactor and the coolant, coolant flow, and coolant valve installed characteristic.

Cascade control of reactor temperature to coolant temperature makes the process gain simple and constant and corrects for coolant disturbances and a nonlinear or non-ideal valve response before they appreciably affect reactor temperature.

6-2. Disturbances and Difficulties

The biggest disturbance is often the set point change for batch reactors or the start-up of continuous reactors. Set point profiles that exponentially increase the set point at the maximum possible rate of temperature rise without overshoot are effective if the bending over of the set point profile is correctly adjusted. This can be simply achieved by a combination of a

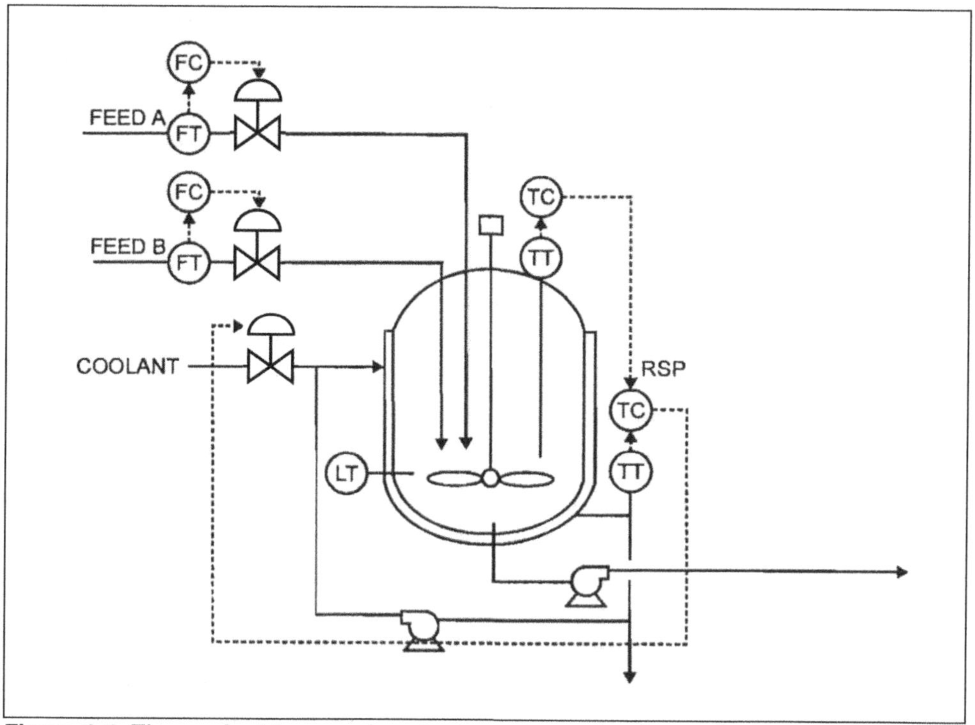

Figure 6-1. The coolant temperature loop compensates for coolant upsets before they affect the reactor temperature and makes the process gain more linear for the reactor loop

ramp and filter or a velocity-limited filter. A ramp by itself has too sharp a corner at the final set point to prevent overshoot. An alternative, particularly effective for following an optimum batch profile, is the translation of the controlled variable from temperature to rate of change of temperature. In terms of the effect on the original controlled variable, the modes are differentiated. Thus, gain becomes derivative, integral becomes gain, and derivative becomes second derivative. This translation is particularly effective for suppressing overshoot and protecting against runaway reactions [1]. The fastest possible response to a step change in set point can be achieved by smart bang-bang logic (full throttle) as described in Section 3-5.

The most frequent load disturbance is a change in makeup coolant temperature or pressure. Cascade control, as previously mentioned, can effectively mitigate the consequences of coolant upsets. Of course, it is always best to correct the source of the disturbance and improve the coolant system design and control. Some deficiencies cannot be corrected by cascade control. For example, the coolant supply temperature may be too warm during hot summer days for the rated or desired expanded capacity.

In some cases, action of the coolant temperature control system is the source of an upset. This occurs for split range systems. Inevitably, an oscillation in coolant temperature can be traced to a crisscross of the split range point. The transition from chilled water or cooling tower water to steam is the worst, especially if it occurs directly in a reactor jacket or coil instead of indirectly in a tempered water heat exchanger. To reduce the propensity to oscillate, the split ranging should be done in the distributed control system (DCS) or programmable logic controller (PLC) rather than in the field positioner or current-to-pneumatic (I/P) transducer; the split range point should be chosen or the controller gain scheduled to compensate for gain changes. If possible, the controller output limit should be set to prevent an excursion onto the wrong valve for a given mode of operation. For example, during heatup or endothermic operation, an output limit should prevent stroking of the chilled water valve, and, during exothermic operation, an output limit should be set to prevent stroking of the steam valve.

The use of a split range gap and the enhanced PID in Figure 9-26 can eliminate the limit cycle from the discontinuity at the split range point to stop the crisscross of the split range point. A small minimum opening of the coolant valve for low steam demand may waste less energy besides provide tighter temperature control by eliminating oscillations from steam shock. In this case, the residence time from the injection of steam to the temperature sensor should have a residence time of at least 0.2 seconds to ensure that no water droplets form as noted in Section 1-6 for desuperheaters.

The oscillation of temperature loops across the split range point that wastes energy can be mitigated by an enhanced PID and a minimum coolant flow for low steam demand.

The most difficult nonlinearity to correct for is a change in heat transfer coefficient due to coating or fouling of the heat transfer surfaces. The bulk fluid velocity past the heat transfer surfaces for even highly agitated reactors rarely exceeds more than one foot per second, whereas seven feet per second are needed to reduce the onset of coating or fouling. The decrease in the heat transfer coefficient causes an increase in the process time constant, which requires a decrease in the controller gain when jacket temperature is manipulated. The result is sluggish control if the controller is not retuned and a possible runaway response for an exothermic reaction. The heat transfer coefficient for constant coolant flow can be computed by taking the difference between the outlet and a delayed inlet temperatures, multiplying the difference by the coolant flow rate to get the heat transfer rate, and dividing the rate by the log mean temperature difference to get the product of the heat transfer coefficient and area. This can be used for adaptive tuning or gain scheduling, as outlined in Section 6-4. Note that the inlet temperature is passed through a dead time block to account for the coolant transportation delay [2]. If the coolant flow rate changes, the calculation becomes much more difficult due to the change in coolant delay and heat transfer coefficient with flow. In this case, a filtered periodic calculation of the rate of change of the heat transfer coefficient with respect to flow might be more instructive as a detection of the onset. In any case, the intermediate calculation of heat transfer rate is also valuable as an indicator of conversion. The heat transfer rate or total can be used for end point control of batch reactors.

The reactor feed composition and consistency are usually well maintained, except when the storage or feed tanks are too small, the flowmeters are too noisy, the feed control valves have stick-slip, the feed controllers are poorly tuned, an upstream unit operation is erratic, or a recycle stream is used. Of these upsets, the variable composition of recycle streams causes the most problems. Online analysis of recycle streams, although expensive, is often justified upon scrutiny of the possible uncertainties and associated penalties. Numerous studies have revealed large capacity and yield improvements from the use of online analyzers to provide actual or online models to create inferential measurements of recycle composition.

6-3. Loop Performance

Reactors determine the quantity of products and byproducts produced. There is often an optimum temperature for conversion (amount that is reacted) and selectivity (amount that is product versus byproduct). Thus, reactors largely determine the yield and waste of a production unit. High temperatures may increase conversion but start side reactions that increase the concentration of undesirable byproducts, many of which are often unidentified. Low temperatures increase the concentration of unreacted feeds. An improvement in reactor temperature control can mean a reduction in raw material consumption and waste production, which makes it important from the viewpoints of competitiveness and the environment. Reactors are usually much smaller users of energy than columns or furnaces, but reactor operation can waste energy from oscillations in the coolant temperature, particularly around a split range point.

6-4. Controller Tuning

The process gain of the coolant temperature control loop (see Equation F-9a in Appendix F) can be approximated as the change in coolant temperature divided by the coolant flow rate. For a constant coolant flow, this means the process gain is proportional, and, consequently, the controller gain is inversely proportional to the temperature difference or driving force.

Considerable improvement in reactor control systems from adaptive control has been documented. The combination of a decrease in coolant controller gain proportional to an increase in coolant driving force, and an increase in reactor controller gain for an increase in deviation of reactor temperature from set point, reduced temperature excursions as large as 20° to 0.5°F in polymer reactors. It essentially eliminated the three million pounds of offspec product and the associated economic penalty (in 1981) of five to ten cents per pound [3]. The improvement from the increase in reactor temperature controller gain with error is consistent with the findings on the benefit of such nonlinear action by fuzzy logic controllers for temperature control loops in general.

Equations F-6h through F-6k in Appendix F provide a more detailed look at the process gain for the reactor temperature controller when its output is used to manipulate either coolant temperature, feed rate, and feed temperature. These equations share a common denominator (c) with the process time constant (compare Equation F-6g and F-6h in Appendix F). As a result, the process time constant can be viewed as proportional to the process gain multiplied by the ratio of numerators. Interestingly enough, for manipulation of coolant temperature, this ratio is the product of mass and heat capacity divided by the product of heat transfer coefficient and area, as shown in Equation 6-1. It is the same relationship that is the thermal time constant for jackets and coils, as discussed in Section 6-1, and thermowells. It follows, from the definition of an integrator gain being the process gain divided by the process time constant, that it is also the inverse of the integrator gain for a zero discharge flow, which occurs for storage tanks and batch operation as depicted by Equation F-6i.

$$\tau_p = (M_r * C_r)/(U * A) \qquad (6\text{-}1)$$

where:

A = heat transfer area (ft.)

C = reactants heat capacity (Btu/lb * °F)

M = reactants mass (lb)

τ_p = process time constant for the reactor loop (min)

U = overall heat transfer coefficient (Btu/min*ft)

For a process time constant much larger than the loop dead time, which is normally the case for the reactor temperature loop, the controller gain is proportional to the process time constant divided by the process gain, per Equation 3-18. This means the reactor controller gain is proportional to the ratio. For cascade control of reactor to coolant temperature (the normal case), the reactor controller gain is proportional to the reactant mass and inversely proportional to the heat transfer coefficient and area. If a change in reactor level equally changes both the fractional volume and heat transfer surface covered, the controller gain is unaffected. If an increase in level increases the mass more than the heat transfer surface coverage, the controller gain can be increased. This would occur for a level change above the coolant coil. The converse is true.

Properly designed reactor temperature loops can have much higher controller gains and much tighter control because the process time constant is so much larger than the loop dead time.

For exothermic reactors, a positive feedback condition can develop because the reaction rate, as expressed by Equation 6-2 is an exponential function of temperature. Note that the reaction coefficient is normally a very large number (e.g., $a = e^{30}$). A runaway condition occurs when the heat removal rate (the first two terms in the denominator) is smaller than the heat generation rate (the last term in the denominator for the process time constant). The result is a negative denominator, a reversal of the sign from negative to positive of the feedback term in Equation F-6g and, consequently, a positive feedback time constant. Without sufficient feedback correction, the temperature will diverge from set point until the heat removal rate equals the heat generation rate. As shown in Figure 6-2, the unstable region occurs when the slope of the S-shaped heat evolution curve exceeds that of the heat transfer line. Inadequate feedback controller action will result in the temperature cycling between the lower and upper intersection points of the curve and line [4].

$$k = a * e^{-E/(R * T)} \quad (6\text{-}2)$$

where:

a = reaction constant (min^{-1})

E = activation energy (Btu/lbmol)

R = universal gas constant (1.987 Btu/lbmol * °R)

T = absolute temperature (°R)

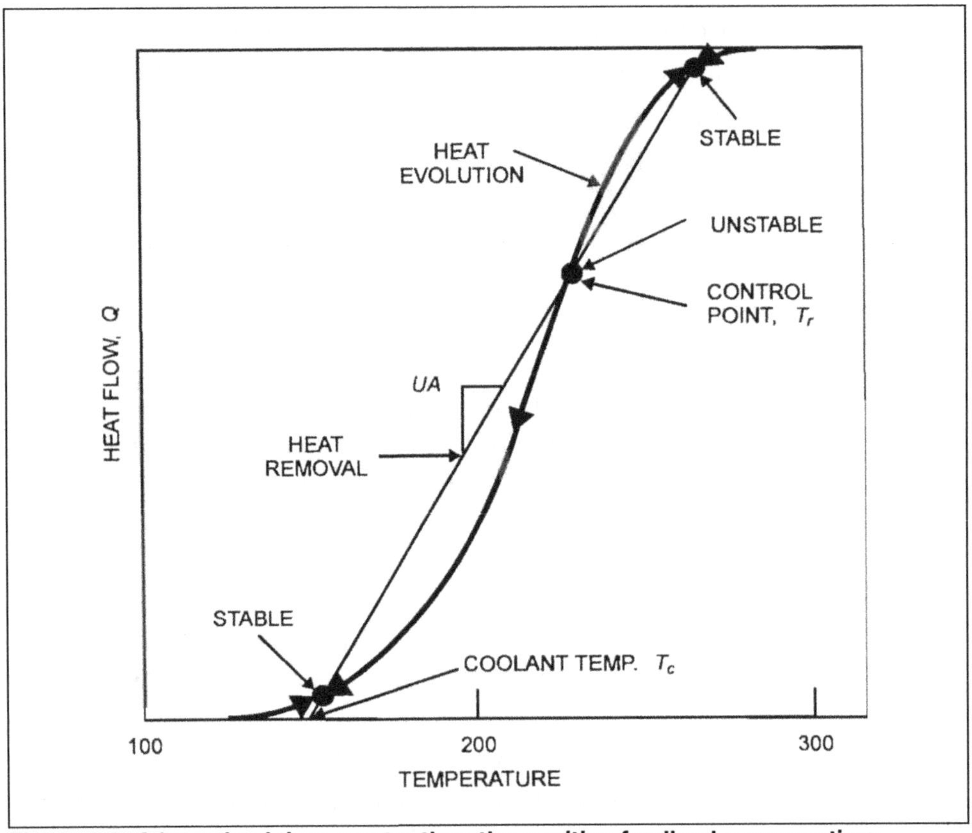

Figure 6-2. A loop dead time greater than the positive feedback process time constant causes a sawtooth cycle between the lower and upper intersection points of heat evolution and transfer [4]

Inadequate temperature controller action for exothermic reactors occurs whenever the controller gain is lower than the inverse of the open-loop gain. thus, a gain window exists where two little as well as too much gain can cause instability. If the loop dead time or any secondary time constant approaches the major process (positive feedback) time constant in size, the window of allowable gains closes, and the loop is unstable for all tuning settings [5]. The size of the window or ratio of maximum to minimum controller gains is proportional to the time constant to dead time ratio for well-designed reactors. The ratio should be at least ten to allow for changes in loop dynamics and gains. Thus, the loop dead time and any secondary loop time constant should be less than one-tenth of the positive feedback time constant.

The process deadtime and any secondary process time constant must be less than 1/10 the positive feedback time constant to allow a window of controller gains large enough to prevent a runaway response for changes in the process gain.

If the reactant feed is manipulated for reactor temperature control, there is an additional time constant for the composition response in series with the major process time constant. Its value, per Equation 6-3, is quite large except for a small residence time (i.e., small V/F) and a slow reaction rate (i.e., a large k). Since reactor residence time is usually large and the reaction kinetics quite fast, considerable additional dead time is added to the loop. Also there is an inverse response for operating temperatures above the feed temperature, because the initial effect of an increase in feed flow is a decrease in reactant temperature as the cool feed mixes with the hot reactants [6]. The drop in temperature from the decrease in sensible heat from the feed is followed by a rise in temperature from the heat of reaction. The initial response, which is opposite of the final response, confuses the controller and necessitates the reduction of gain and derivative (rate) settings, which is detrimental to all reactors and potentially disastrous for exothermic reactors. The advantage of this loop arrangement is a natural maximization of reactor capacity because the feeds are increased to the capability of the heat removal system.

The major process time constant for direct manipulation of feed is the result of backmixing, which distributes the heat changes that originate in the reactor volume rather than heat flow across heat transfer surfaces (e.g., coil or jacket). Thus, the heat transfer coefficient and surface don't enter into the analysis, and Equation 6-1 doesn't apply. For backmixing of fluids at different temperatures, the major process time constant for a change in feed, is the residence time (Equation F-7g) minus the mixing dead time (turnover time) (Equation F-6m).

Most of the capacity improvement from direct manipulation of the feed can be achieved, as depicted in Figure 6-3, without the creation of inverse response or the addition of dead time by the use of a valve position controller that gradually maximizes reactor throughput. The valve position controller seeks to keep the coolant valve position, which is the controlled variable, at a maximum controllable value (e.g., 80%) by manipulation of the reactant feed rate. While valve position controllers normally use just integral-only action for slow optimization, the other modes and an error squared algorithm or feedforward control can be used to provide a more responsive feed correction to prevent a wide-open coolant valve and temporary loss of feedback action for big upsets. However, gain, reset, and rate action of the valve position controller must be much less than that used by the reactor temperature controller to prevent overreaction to the inverse response and interaction between loops.

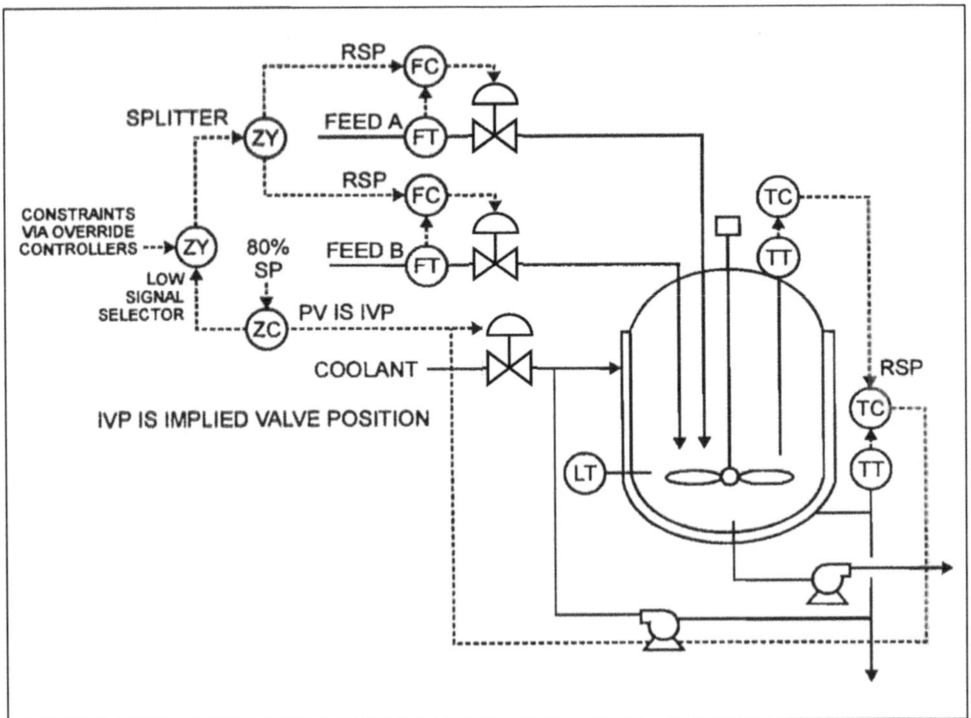

Figure 6-3. A valve position controller can optimize production rate by maximizing coolant valve position

$$\tau_x = V_r/(F_o + V_r * k) \tag{6-3}$$

where:

F_o = flow rate out of the reactor (gpm)

k = reaction rate (min)

τ_x = composition time constant from manipulation of feed (min)

V_r = reactant volume (gal)

Since the heat released by reaction is equal to the heat of reaction (H_r) multiplied by the incoming controlling reactant feed flow ($F_f * X_i$) that is converted (Y), the change in heat evolution (dQr/dT) with temperature in the denominator (c) in Appendix F for the process gain and process time constant depends, as shown in Equation 6-4 upon the change in conversion with temperature (dY/dT) at constant concentration (X). The differentiation of the combination of Equations 6-2 and 6-3 at constant concentration for a backmixed reactor inserted into Equation 6-4 gives Equation 6-5 for the change in heat release with temperature [6]. This equation applies to backmixed reactors for manipulation of either coolant temperature or feed flow for temperature control.

$$dQr/dT = H_r * F_f * X_i * (dY/dT)_x \tag{6-4}$$

$$dQr/dT = H_r * F_f * X_i * (Y * E)/(R * T^2) \tag{6-5}$$

where:

E = activation energy (Btu/lbmol)

F_f = feed flow (lb/hr)

H_r = heat of reaction (Btu/lb)

dQr/dT = change in heat evolution with temperature (Btu/°F)

R = universal gas constant (1.987 Btu/lbmol * °R)

T = absolute temperature (°R)

X_i = fraction of incoming controlling reactant in feed

Y = fraction of incoming controlling reactant converted

$(dY/dT)_x$ = change in conversion with temperature (1/°F)

The use of Equation 6-5 with knowledge of the overall heat transfer coefficient and area and the reactor operating conditions allows computation of the process gain and process time constant, which when combined with calculations of the secondary time constants and pure dead times, can provide information on loop dynamics and the controller tuning settings. For exothermic reactors, the reset action should be reduced, because the positive feedback of reset action combined with the positive feedback in the process is particularly disastrous. In studies of the tuning for a runaway reactor, it was found that to eliminate reset induced overshoot and oscillations, the reset time needed to be ten times larger than for a conventional batch temperature loop. In applications where the threat of a runaway is a concern, reset action is omitted, and a proportional-plus-derivative (PD) controller is used.

The controller gain, reset time, and rate time must all be maximized to prevent a runaway reaction.

However, there may not be enough data or time to go to this much effort. In fact, the main use of these equations is to show relationships that are important for improving reactor control. The tuning settings shown in Table 6-1 for temperature control of an exothermic reactor by manipulation of coolant temperature are useful as a shortcut estimate and as a check of calculations. Actual or computed tuning settings outside of the range means that you have either made a mistake or you have an unusual exothermic reactor and/or instrumentation system (e.g., excessive loop dead time or extremely small temperature calibration spans).

Table 6-1. Typical tuning settings for an exothermic reactor temperature controller use high gain and derivative and low or no reset action to prevent the possibility of a runaway reaction

Gain	Reset	Rate
5.0 to 50.0	600 to 6000 seconds	60 to 600 seconds

Since the general tendency is to use too much reset action and not enough gain or rate action, the performance of nearly all types of reactor temperature loops can be greatly improved by reducing the reset action by a factor of 10 or more, by doubling the gain (after the reduction in reset action) and using at least 0.5 minute of rate. Again, it is critical to remember that a decrease in reset action is achieved by a decrease in the repeats/minute and by an increase in the minutes per repeat.

6-5. Control Errors

The peak error (i.e., maximum deviation from set point) of a well designed, installed, and tuned reactor control system is potentially a lot smaller than that for most other types of temperature loops. The peak error is usually less than the measurement error. While most of the measurement error is a bias or offset that could be corrected by an adjustment of the controller set point or by a zero adjustment of the transmitter, the value is often unknown. Also, the A/D and high-order polynomial noise of linearizers of wide calibration spans prevent the use of large gain and rate settings. Thus, most reactor control systems benefit from the use of RTDs and narrow range transmitters, an old practice somehow forgotten with the advent of the DCS and PLC and their associated thermocouple and RTD input cards.

6-6. Control Strategies

Table 6-2 summarizes the relative advantages and disadvantages for different pairings of controlled and manipulated variables.

These control strategies are grouped for comparison by method of heat removal or supply and operational mode. The first four strategies apply where coolant is used for heat removal; the next two apply where steam is used as a source of heat; the subsequent two apply where a vent or vent condenser is used for heat removal; and the last strategy applies to set point changes.

For removal of heat, coolant temperature, as previously discussed, is the best manipulated variable, because it compensates for changes in coolant upsets before they affect the reactor temperature and helps linearize the process gain for the reactor loop. The choice of whether to use inlet or outlet temperature depends upon whether the main source of upsets originates from the coolant system or from changes in the heat transfer coefficient. the inlet coolant temperature loop has the least loop dead time due to the short distance between the inlet temperature and the manipulated blending of the makeup flow with the recirculation flow. The outlet coolant temperature responds to a coolant upset after the transportation delay from plug flow through the coil and/or jacket. Similarly, the inlet temperature responds to a heat transfer coefficient upset after the transportation delay from plug flow in the recirculation piping. For high coolant flow rates, the additional time delay is not a factor, because it is much smaller than reactor temperature loop dead time or process time constant, and either coolant loop temperature performs well for both types of disturbances. The performance advantage of cascade control of reactor temperature to coolant temperature more than justifies its use, even though the direct throttling of makeup coolant flow is slightly less expensive and complex. The throttling of total coolant flow is not listed, because it is not considered a safe method of heat removal for exothermic reactor temperature control.

Table 6-2. The strategies with the best performance use coolant temperature or steam pressure as an inner loop for the main reactor control, and a computed equilibrium temperature or a rate of change of temperature as the controlled variable, for auxiliary vent/reflux control and set point changes, respectively

Controlled Variable	Manipulated Variable	Relative Advantages	Relative Disadvantages
Reactor temperature	Inlet coolant temperature	Faster response to coolant temperature and pressure upsets	Slower compensation for fouled or coated heat transfer surface
Reactor temperature	Outlet coolant temperature	Faster compensation for fouled or coated heat transfer surface	Slower response to coolant temperature and pressure upsets
Reactor temperature	Coolant makeup valve	Less expensive and complex installation	More coolant upsets and nonlinearity
Reactor temperature	Reactant feed flow	Higher capacity (throughput)	Inverse response and more loop dead time
Reactor temperature	Steam pressure	Less heat load upsets and nonlinearity	No interference of energy use (Btu/hr)
Reactor temperature	Steam flow	Inference of energy use (Btu/hr)	More heat load upsets and nonlinearity
Reactor temperature	Reactor pressure	Faster response and less foam and vent loss	No built-in use of equilibrium relationship
Computed reactor temperature from pressure	Reactor vent valve	Fastest response and least foam and vent loss	Slow correction of equilibrium relationship
Computed reactor temperature rate of change	Coolant temperature of steam pressure	Less overshoot and smoother approach	Unusual strategy and translation of tuning mode

For the use of steam as a heat source, steam pressure, as discussed in Section 5-1 for heat exchangers, will automatically correct for changes in condensing rates and thus provide self-compensation for changes in heat loads. The use of steam flow as a manipulated variable provides a measure of the energy usage. The use of a steam flow differential head meter will keep the energy input constant for steam pressure changes. However, the general trend is to eliminate orifice flowmeters and the hidden installation, downtime, and service costs of keeping sensing lines full of condensate without plugging or freezing. For large steam lines and insertion-type flowmeters, erratic readings from insufficient straight runs are of more than normal concern.

The use of differential condenser temperature control of vents, valves, and vent condensers, to remove reactor heat by vaporization and condensation, develops a limit cycle that causes excessive product loss out the vent and fouling of vent lines, condensers, and knock outputs. During high pressure excursions, the layer of bubbles extends deeper into the reactant mixture, which causes more foaming (similar to a covered pot of boiling water and spaghetti). During low pressure excursions, the foam surges into the vent system and causes short and long term changes in the heat transfer coefficient due to bubbles and fouling. The result is more downtime due to cleanouts and greater product loss out the vent.

The use of vapor velocity control can mitigate some of this, but the best solution is pressure control. This can be achieved by cascade control of reactor temperature to vapor space pressure. However, the equilibrium relationship between temperature and pressure can be used to compute a temperature from pressure that provides even faster and tighter temperature and pressure control. A simple linear relationship, as defined by Equation 6-6, has been found to be best rather than more complicated expressions that attempt to include the effects of component vapor pressures. To compensate for changes in the relationship with composition or due to measurement inaccuracies, an integral-only controller whose measurement is the computed temperature and whose set point is the actual temperature is used to adjust the bias (B) [7]. The process variable filter for this adaptive controller is set equal to the lag between the actual and the computed measurements. The slope or gain coefficient (A) can be preprogrammed based on batch or start-up time or heat removed as an inference of composition changes proportional to conversion. Since the computed P vs. T loop is so fast, there is no interaction between this and other loops to control reactor temperature via the coolant system. Figure 6-3 shows how a valve position loop can be added to minimize vent losses by shifting as much of the heat removal load to the coolant system by a valve position controller that seeks to move the vent valve to a minimum controllable position (e.g., 10%).

$$T = A * P + B \tag{6-6}$$

where:

A = slope or gain coefficient (°F/psi)

B = bias (°F)

P = vapor space or condenser pressure (psig)

T = computed reactor equilibrium temperature (°F)

The use of temperature rate of change as the controlled variable will provide a smooth consistent approach to a set point without overshoot. The reactor temperature is filtered and passed through a dead time block to create an old value from which a new value is subtracted. The dead time is chosen to make the temperature difference large enough so that A/D noise is a negligible fraction of the real change. The differentiation of the tuning modes with respect to the original controlled variable, means there is no reset action to remove offset. Figure 6-4 depicts the low signal selection of the outputs from the rate of change controller and a conventional PID temperature controller to eliminate offsets above set point for a high exothermic reactor.

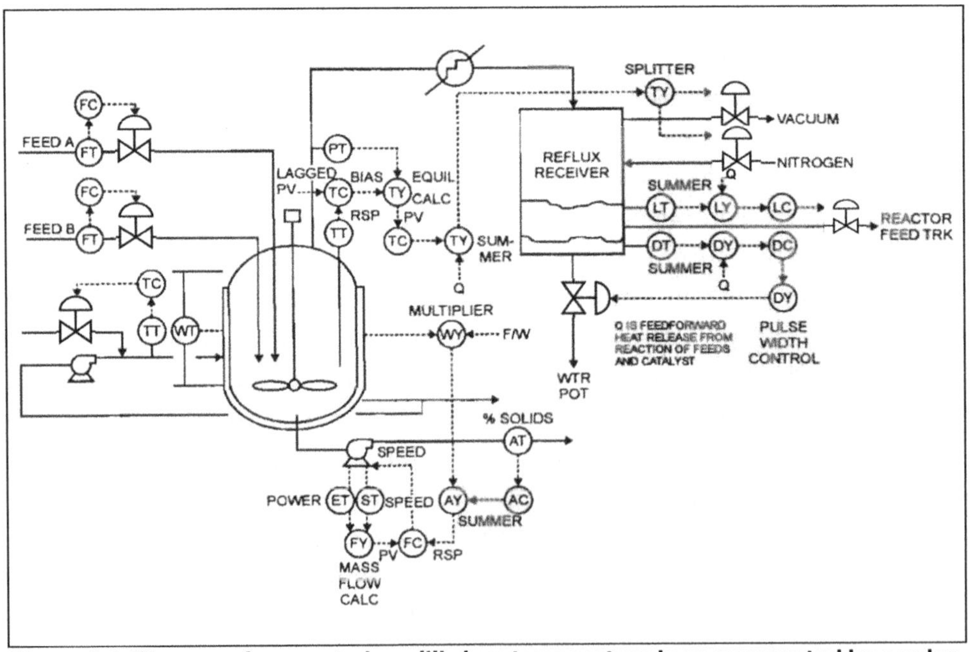

Figure 6-4. The use of computed equilibrium temperature loop augmented by a valve position controller can minimize product vented and fouling of the vent system

To maximize reactor production, whether batch or continuous, the feed rates can be pushed to the limit of identified constraints by either model predictive control or by override control. Figure 6-4 illustrates how a wide variety of constraints, including reactor liquid temperatures, vapor and liquid differential temperatures, and a tails tower condenser temperature, are used as set points of override controllers whose output passes through a low signal selector to manipulate the primary reactant feed rate. Note that there is a total flow controller for multiple injection streams of the primary reactant, rate changes are velocity limited, and secondary reactants are ratioed to make smooth transitions in production rate. The override controllers are proportional-plus-rate controllers, so external feedback signals and the filters for stability that add delay are not needed. Also, the omission of rest action means the relationship between override controller output and distance to a constraint is defined and controlled by the override controller bias.

Model predictive control and override control and maximize reactor feed to ride the constraints of maximum coolant, condenser, and vent valve position.

The key variables for supervisory control are (1) conversion (pounds of reactant used divided by the pounds of reactant fed) and (2) yield or selectivity (pounds of product formed divided by pounds of reactant used).

Total heat removal for exothermic and heat supply for endothermic reactions can be an inference of conversion. The temperature measurements must be extremely accurate for small temperature differences and compensated for liquid holdup. Also, the sensible heat contribution from the net temperature difference between the feeds and reactor contents must be backed out of the calculation.

Portable wireless integral mount RTD temperature transmitters can be used to test the viability of a heat removal calculation for an inference of reaction rate. The inlet temperature measurement used in the calculation is delayed by a dead time that is the jacket or coil volume divided by the coolant flow.

Direct conversion and yield calculations require composition measurements from online analyzers or "soft sensors," also known as "intelligent

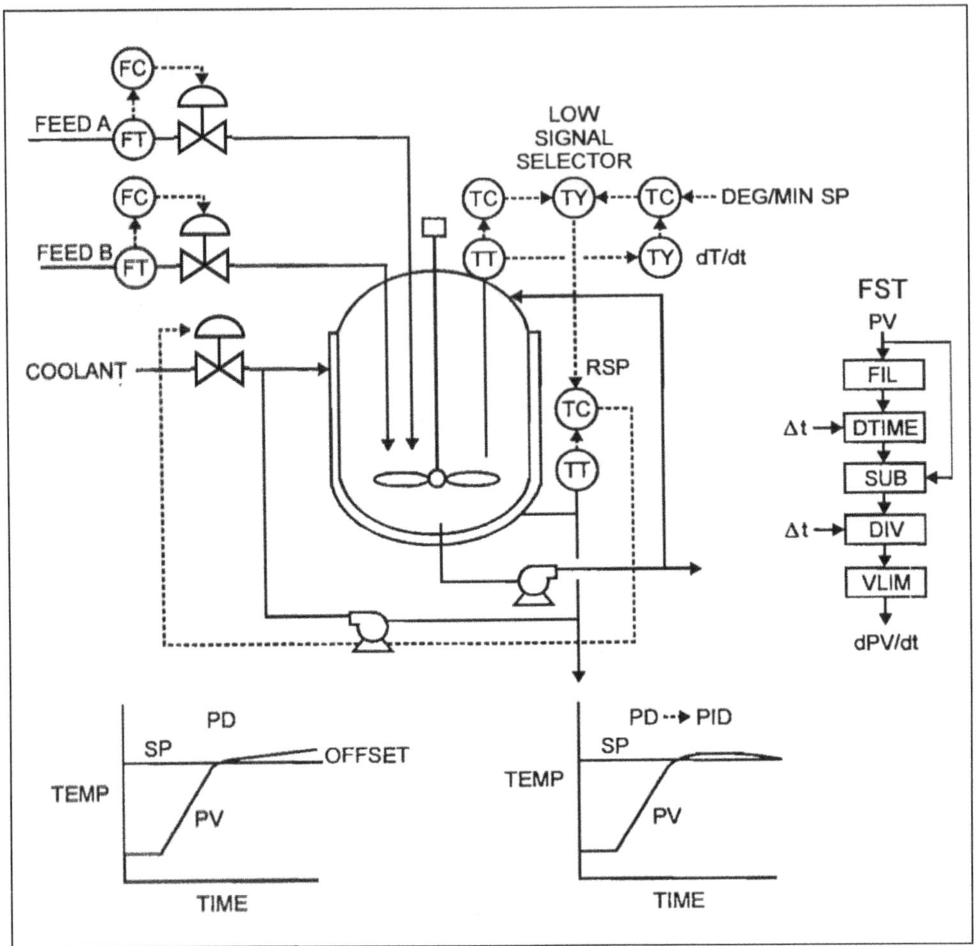

Figure 6-5. A log output signal selection for coolant temperature set point of the rate of change and conventional temperature controller outputs can provide a smooth approach with no overshoot and no offset above set point

sensors" (software calculations of component concentrations from online dynamic models or neural networks).

The choices of manipulated variables for supervisory control are reactor temperature, residence time, and mixing. Higher reactor temperature and residence time generally increase conversion but may decrease yield through byproduct formation. Every ten degrees centigrade corresponds to a doubling of the reaction rate. Betting mixing can increase yield and conversion and reduce the formation of clumps or chunks. However, mixing can cause excessive heat, breakage of crystals, and rupture of cells.

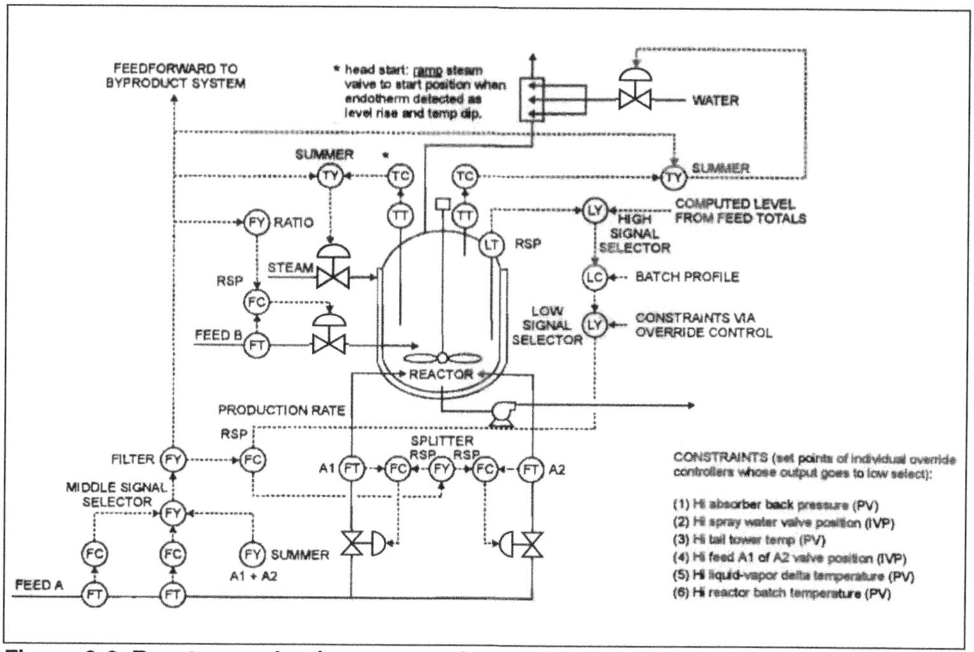

Figure 6-6. Reactor production rate can be maximized by pushing against constraints enforced by simple proportional-derivative plus bias (PD + bias) override controllers

Exercises

6-1. What is the most difficult disturbance to correct for while controlling reactor temperature?

6-2. What is the most common way to correct for coolant temperature excursions while controlling reactor temperature?

6-3. For exothermic reactors, there is a window of allowable controller gain for temperature loops. What should be the ratio of the minimum stable gain setting to the maximum stable gain setting, and why is there a temperature controller gain window for stability?

6-4. What should be the reset setting on the reactor temperature loop where there is the threat of a runaway reaction?

6-5. Reactor temperature is to be controlled via heat removal. Should outlet or inlet coolant temperature control be used?

References

1. McMillan, G. K., *Process Control Case Histories: An Insightful and Humorous Perspective from the Control Room*, Momentum Press, 2010.
2. Nowicki, P. L., "Improve Reactor Control," *Chemical Engineering Progress*, December 1991.
3. Whatley, M. J., and Pott, D. C., "Adaptive Gain Improves Reactor Control," *Hydrocarbon Processing*, May 1984.
4. Shinskey, F. G., "Uncontrollable Processes and What to Do About Them," *Hydrocarbon Processing*, November 1983.
5. McMillan, G. K., *Tuning and Control Loop Performance*, 3rd edition, ISA, 1994.
6. Shinskey, F. G., *Process Control Systems*, 3rd edition, McGraw-Hill Publishing Company, 1988.
7. Hopkins, B., and Alford, G. H., "Temperature Control of Polymerization Reactors," *Instrumentation Technology*, May 1973.

7
Columns

This chapter describes control strategies and operational considerations for distillation columns.

Learning Objectives

A. **Understand how operation close to the weep or flood point will disable control systems.**

B. **Know how to compensate for large diameter distillate receivers.**

C. **Determine the best plate for temperature control.**

D. **Be able to construct the best control strategy.**

7-1. Process and Equipment Design Considerations

Column design that causes operation at the weep point or flood point will prevent stable operation. The operating point can be moved further away from the weep and flood points by a decrease and an increase in column pressure, respectively. It is desirable to reduce the pressure set point, if possible, because the relative volatility, which is a measure of the ability of the column to separate components, improves as the pressure is lowered. Most columns can run at full vacuum if the operating point is far enough away from the flood point and the condenser has sufficient heat removal capability (duty) for the higher vapor flow.

The column pressure should be as low as possible to reduce energy use but must prevent weeping or flooding that would disable temperature control and cause column instability.

When a column goes into weep, the vapor flow is insufficient to keep the liquid suspended on the trays. Liquid falls through the holes in the trays and can eventually end up mostly in the column sump, along with some trays. Temperature sensors in the upper portion of the column run dry. The pressure drop plummets and the sump level rises from the dumping.

When a column goes into flood, the vapor flow is so high it causes the froth from one tray to intermittently choke and break through to the next tray. The result is a percolation, much like in a coffee pot, that can cause spillover into the distillate receiver. The pressure rapidly fluctuates and the receiver inventory may surge. For large horizontal drums, the level change might not be noticeable due to the large cross-sectional area. The column temperature profile can become inverted as high boilers are burped upward and low boilers are swept downward.

Since the use of the best tray location is critical and can change with operating conditions, it is extremely important that the column have temperature connections at many trays. Sensors should be installed at the best tray and the ones immediately above and below it, as well as at enough other trays to get a good column temperature profile for diagnostics and validation of process models. The sensor should extend into the liquid froth above the tray to maximize the heat transfer coefficient but should not interfere with column traffic. Intrusion of the thermowell into a downcomer area might cause local flooding. In packed columns, sensor location is more critical due to the possibility that channeling will cause liquid to bypass the sensor.

The temperature sensor must extend into the liquid holdup on a tray or in the packing.

The condenser and reboiler duty should permit operation at the lowest and highest column pressure, respectively. Since the coolant pressure and temperature are less well regulated than steam pressure and temperature, the coolant to the condenser is often manipulated to control column pressure, and the steam to the reboiler is set to fix the energy per unit feed.

The distillate receiver diameter should be small enough so that the process integrator gain (ft./lb) is large enough to not require an exceptionally high level controller gain. The level change within the process dead time must be greater than the minimum controllable level and translate to a temperature change that is less than the minimum controllable temperature. The minimum controllable value is set by noise, and the sensitivity and resolution limit of the respective measurements and control valves. Thus, horizontal drums cause problems due to a small and nonlinear process gain. For these applications, either an exceptionally high controller gain (e.g., Kc > 20) or a translation of the control variable from level (%) to inventory (lb) should be used to facilitate tight inventory control. The use of reset action and low gain will create slow, nearly equal amplitude-sustained oscillations. The amplitude increases with reset setting (repeats/minute) and control valve dead band (%) [1].

7-2. Disturbances and Difficulties

The largest disturbance is usually in the feed flow. Feedforward control should be used to help maintain the material balance split via the distillate-to-feed (D/F) ratio, or reflux-to-feed (R/F) ratio, and the column separation via the vapor-to-feed (V/F) ratio to reduce energy (steam) usage by operation closer to product quality constraints. Operation at lower energy per unit ratios and, thus, separations requires tighter control of the split (D/F) [2].

Changes in column pressure will cause changes in column temperature even though the composition is constant. Pressure composition can be used, but it assumes a two-component (binary) mixture. The use of the tray that shows the largest temperature change for a composition change is often the most binary tray and usually exhibits the least pressure effect. Thus, the use of the best tray, as outlined in Section 7-6, eliminates much of the problem directly or indirectly through pressure compensation.

The measurement location with the largest change in temperature for both an increase and decrease in manipulated ratio (e.g., distillate to feed), minimizes the effect of pressure controllability and temperature noise, sensitivity, and resolution.

If the concentration of other than the binary components is not negligible, the inference of composition from the temperature of a boiling mixture deteriorates. Distillation column control, as outlined in this section, assumes a binary mixture. Large and variable concentrations of additional components will require the use of analyzers for composition control.

If the feed already has a low level of product impurity (e.g., ppm), temperature is no longer an inference of composition, and separation has a

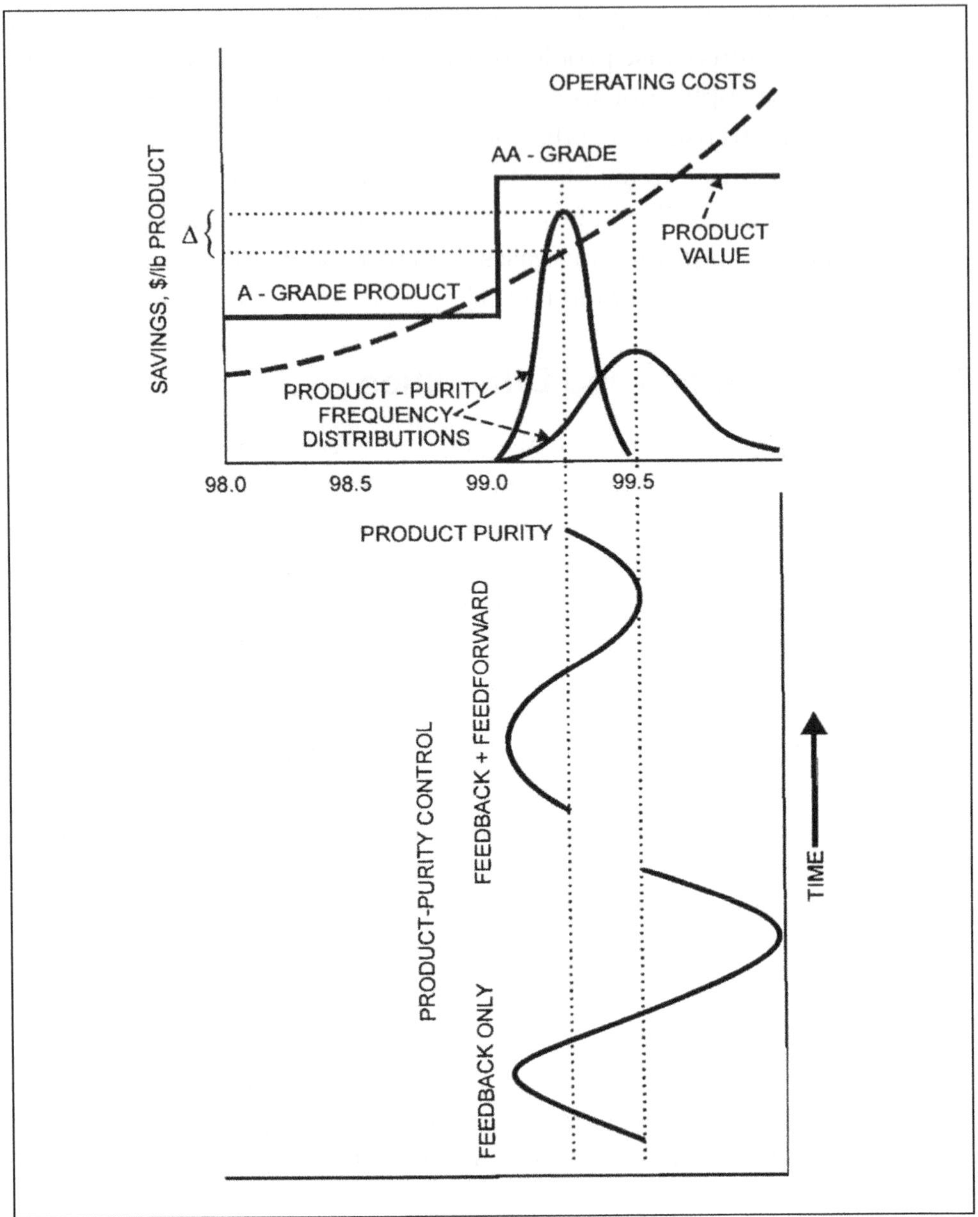

Figure 7-1. Tighter temperature control allows closer operation to a quality constraint that reduces energy consumption [2]

greater than normal effect on composition. In these cases, analyzers need to be added and two-point composition control used to manipulate both the split and the separation. For example, split might be adjusted to control overheads composition and separation might be adjusted to control bottoms composition.

If the concentration of the impurity to be separated is extremely low (e.g., ppm) or the concentrations of more than two components significantly affect temperature, online analyzers and composition control is needed.

7-3. Effect of Loop Performance

Columns are generally the biggest users of steam. The energy savings of several hundred thousands of dollars a year are possible for large recovery and purification systems by the use of better feedback and feedforward control strategies. Also, since these columns are often the last unit operation, the impact on product quality is immediate. Figure 7-1 shows how tighter control means quality can be more consistently maintained while minimizing energy consumption. Besides reduced energy costs, a more narrow product frequency distribution can prevent excursions into lower grade production. A reduction in down-graded product can translate to more revenue from higher product prices and an increased customer base, or less processing and inventory costs for rework or blending.

7-4. Controller Tuning

Except for high purity columns, changes in the split have a ten to one hundred times greater effect on the control plate temperature than changes in the separation. The dynamics for manipulation of split are rather slow due to a time constant associated with each tray's liquid holdup composition response. The time constants are interactive because the adjacent trays mutually affect each other through downcomer flow and vapor flow. This large number of interactive time constants in series establishes, for a first-order approximation, a relatively constant dead time to equivalent time constant ratio of about 0.2. A column usually takes several hours to line out after an upset. For large columns, it may take several shifts. Packed columns tend to have one third the liquid holdup and, thus, are about three times faster than tray columns. The net result, in any case, is that the tuning of a temperature loop that manipulates split requires extreme patience. The chance of a disturbance occurring during the long period while the controller is in manual for an open-loop test is considerable. The closed-loop method is less disrupted by upsets, since the controller is in automatic, but requires about the same amount of time. For a minimum of four oscillations, each with a period of about four times the dead time, the closed-loop test time corresponds to 3 equivalent time constants, which is about the same as waiting for an open-loop test to reach 95% of its final value. The tuning test time can be reduced to about four dead times by the use of the short cut "near integrator" tuning method discussed in Chapter 3.

The tuning can be checked by the identification of the dead time from either the first quarter cycle of a closed-loop test or from the delayed start of the open-loop response. The equivalent time constant is then estimated to be about five times the dead time for the manipulation of split. The open-loop gain is the dimensionless product of the valve (manipulated variable) gain, process gain, and measurement gain. The manipulated variable gain for an approximately linear control valve and for an inner flow loop is estimated as the valve capacity and flow measurement scale, respectively, divided by 100%. The process gain is the change in temperature for a change in split (D/F or B/F ratio) multiplied by the inverse of the feed flow (the change in ratio with respect to a change in distillate or bottoms flow) and can be in ratio with respect to a change in distillate or bottoms flow) and can be obtained from test runs of steady-state simulations. The measurement gain is simply 100% divided by the span of the temperature transmitter calibration or loop scale.

The tuning test time for a large column can be reduced from days to hours by the "near integrator" method significantly reducing the chance of a feed flow or composition change from affecting results.

7-5. Control Errors

The peak error can be rather small due to a relatively good time constant-to-dead time ratio. However, the expectations for column temperature loops are quite high, and improper tuning due to the slowness often yields performance below what is possible. As with most other slow loops, the user tends to use too much reset and not enough rate action, which destabilizes the loop. Often, the controller gain is set too low, but even if it weren't, it would have to be decreased due to improper reset and rate settings.

The integrated error is larger for distillation than for other unit operations due to the exceptionally large magnitude of the dead time and lower than possible controller gains. Thus, the use of good tuning procedures or auto tuners to improve feedback control and the addition of feedforward control can make a huge difference in the control errors.

7-6. Control Strategies

Steady-state process simulations or plant tests can be run to determine the tray with the largest and most symmetrical (linear) response to a change in the split (D/F or R/F ratio). This is also generally the most binary tray and the tray with the fastest dynamic response, as shown by control plate 7 in Figure 7-2 for a change in the D/F ratio. It is also the tray that usually minimizes the maximum excursion of all other trays, as indicated by Figure 7-3 in which the tray temperature extremes are shown for perfect control (pinched temperatures) at control plates 7 and 13 [3]. While other locations (e.g., closer to the end of the column where the flow is manipulated) may provide a smaller dead time, the importance of sufficient process gain (process sensitivity) is paramount. In other words, it does you no good to have a faster loop if the temperature is no longer a good inference of composition. The composition error for a given temperature measurement error can be so large, due to small temperature changes (low process sensitivity), that control of a single tray temperature is meaningless. In some o f these cases, a delta temperature difference between two different tray can be used as the controlled variable to improve the inference of composition.

The allowable change in composition indicated by a corresponding change in split (D/F or R/F), must be less than the minimum controllable temperature set by the temperature measurement and valve sensitivity and resolution.

The manipulation of separation by the adjustment of the energy or vapor per unit feed (V/F ratio) has a dramatically faster response. The process dead time and time constant are seconds rather than hours. Now, the measurement time constant is no longer negligible and may, in fact, be the largest lag in the loop. The manipulation of separation to control composition is not useful except for high purity columns and two-point composition control, because the change in steam and vapor flow required is exceptionally large to get a temperature change. Extreme changes in reboiler steam, which change the column traffic of both vapor and liquid, disrupt column operation and upset the pressure and level loops. Thus, for most columns, temperature is best controlled by the manipulation of split (material balance).

Figure 7-4 shows that the change in light key mole fraction in the overheads (Y) and in the bottoms (X) is small for change in separation (S),

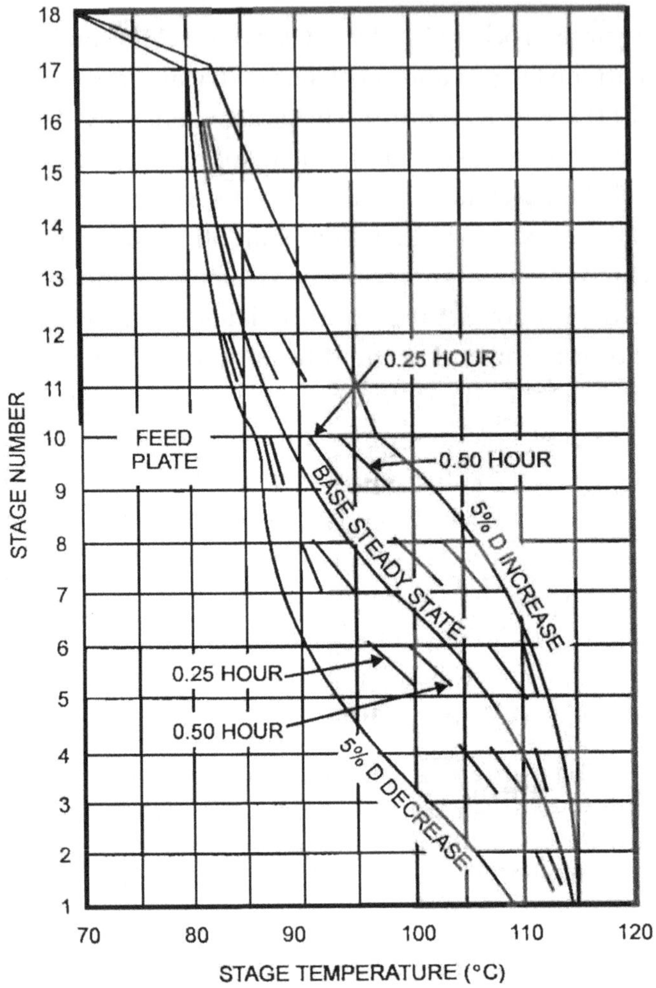

Figure 7-2. The location for the temperature control is the tray with the largest and most symmetrical temperature change [3]

except for distillate to feed (D/F) ratios that are close to zero and one, respectively. The light key mole fraction in the opposite end of the column approaches the light key mole fraction in the feed (Z). Essentially all of the light key component exits the column in the bottoms and overheads, for D/F ratios of zero and one, respectively. The slope of the curve for the light key mole fraction as a function of D/F ratio flattens out at these extremes. Consequently, separation (energy balance) becomes better for manipulation for low and high D/F ratios.

Four major schemes, as depicted in Figures 7-5 through 7-8, for material balance control use column temperature for the inference of tray composi-

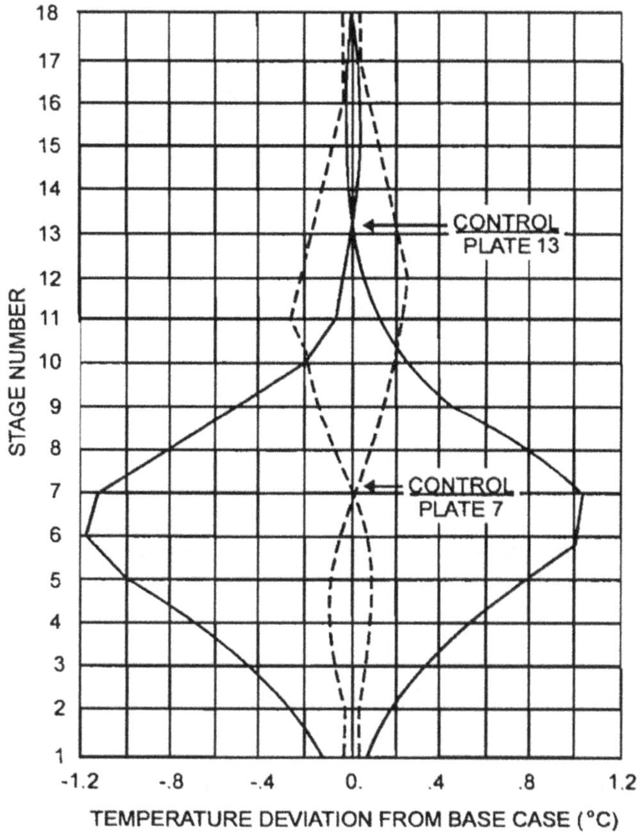

Figure 7-3. The use of the tray with the largest temperature change will minimize the temperature deviation of other trays [3]

tion. Table 7-1 summarizes the relative advantages and disadvantages for each scheme. Where a scheme depends upon another loop besides the temperature loop for composition control by manipulation of the D/F ratio, it is termed indirect material balance control.

The best scheme for most situations is the first because the internal reflux is inherently controlled. If the temperature of the external reflux gets cooler, perhaps due to a rain storm, the lower overhead vapor flow due to cooler reflux translates to a smaller reflux flow from level control in the distillate receiver. The tightness of the level control is key to the success of this scheme, because inferred composition control by temperature control as well as internal reflux control depend upon it. The manipulation of distillate flow for tray temperature control affects the column only through the corresponding change in reflux control via level control of the overheads receiver. Since the ultimate gain of this loop, like most level loops, is

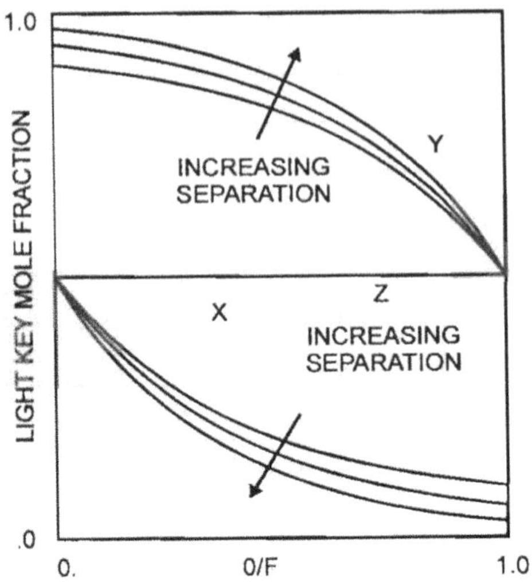

Figure 7-4. The change in composition for a change in D/F ratio is best for column control except for extremely low or high ratios

typically well over a hundred, the highest gain that doesn't cause excessive valve wear should be used with a reset time of 1000 or more seconds and zero rate time. The controller set point and measurement or output should be velocity-limited to prevent spikes in valve position due to set point changes or noise. A filter of up to 0.1 minute can also be used as protection against measurement noise without the introduction of excessive loop dead time. By the use of velocity limits or small filters, the gain should be able to be set larger than fifteen so that there is no additional delay between scheme one and scheme two. Thus, only a personal reluctance to the use of high gains is left as a reason to choose scheme two instead of scheme one.

The manipulation of distillate flow for temperature control compensates for disturbances to internal reflux if the change in level is large enough for the level controller to enforce the material balance.

One of the criteria for selection of the proper scheme is whether the steam manipulated for level control has enough of an effect on level. In other words, a distillate flow much smaller than the reflux flow would dictate the use of scheme two over scheme one, and an extremely small bottoms flow (e.g., purge flow) would necessitate the use of scheme three instead of scheme four. In the last instance, control of level by boil-up can be

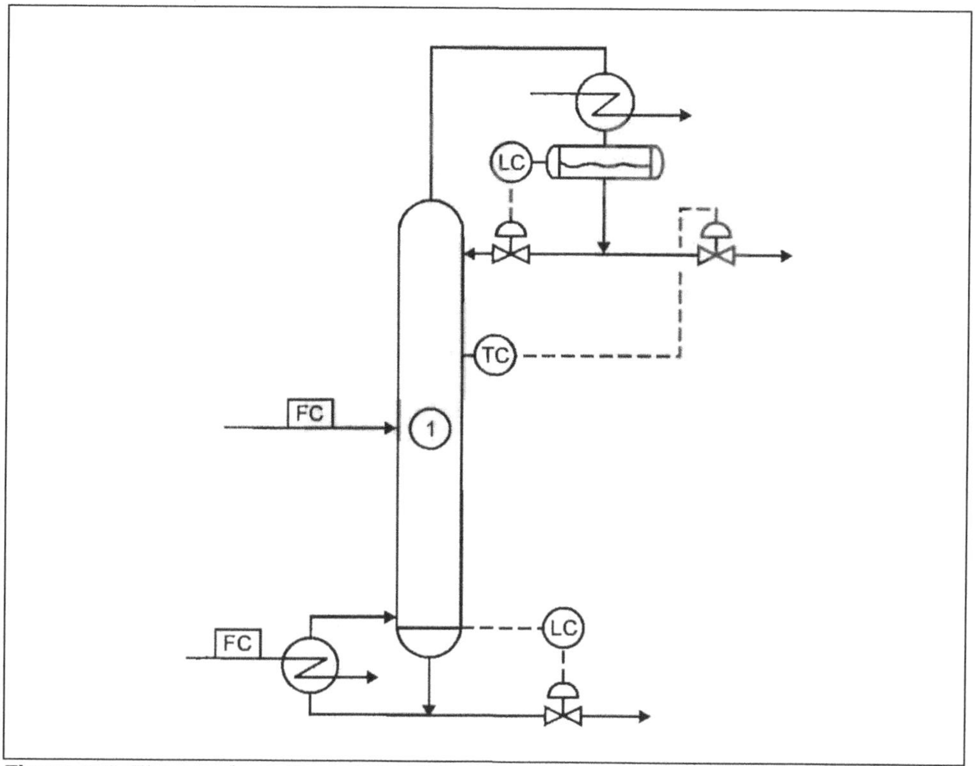

Figure 7-5. The manipulation of distillate gives internal reflux control and minimizes interaction between the material and energy balances but requires high overheads level controller gains [3]

exceptionally difficult due to inverse response. A model predictive controller that recognizes the self-inflicted irregular level response may be helpful. Feedforward control where the steam flow to feed ratio is slowly corrected by level control provides preemptive correction and mitigates the effect of inverse response.

The manipulation of reboiler steam to control temperature causes the greatest interaction between the material and energy balances. The interaction is less for applications in which the effect of separation changes on composition control are smaller. The enforcement of the material balance depends upon the sump level controller manipulation of bottoms flow. Thus, a high level controller gain is desirable. However, besides valve wear, the disturbance to controls on downstream equipment must be considered. Also, shrink and swell, due to bubbles in the return from the reboiler, forces detuning of the level controller. Feedforward signals are often needed to maintain the material balance and/or smooth out the response to reduce upsets to downstream loops.

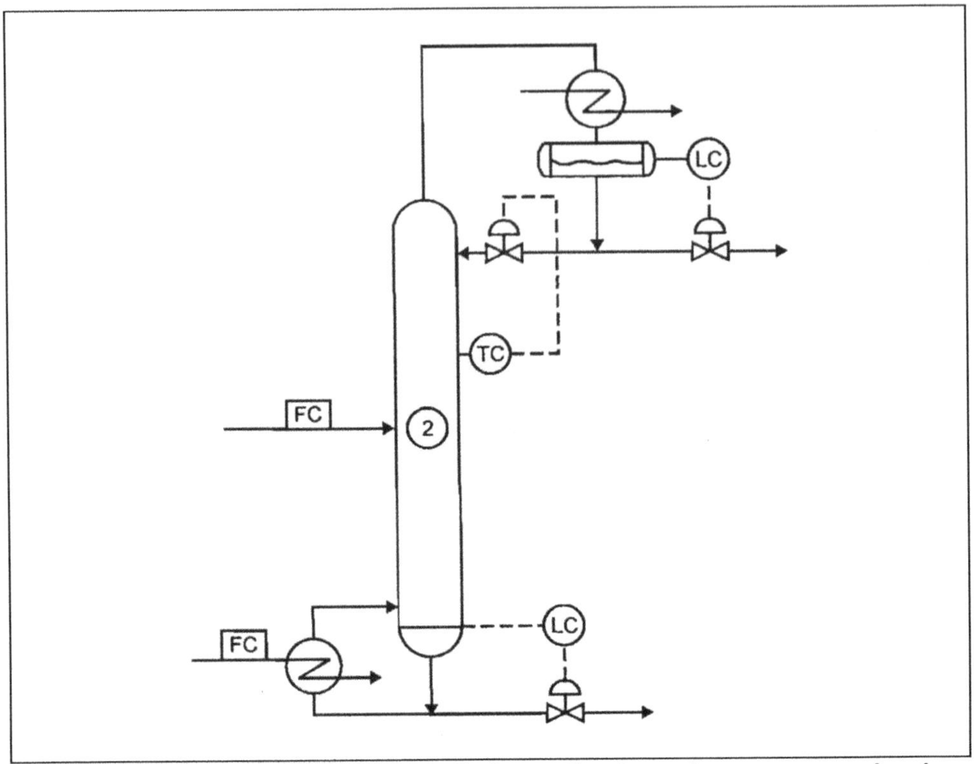

Figure 7-6. The manipulation of reflux eliminates the dependence upon overheads level control for composition control of the column but causes interaction between the material and energy balances [3]

Feedforward flow control of sump level provides the fastest and tightest enforcement of material balance by preemptive action and minimizes the effect of inverse response.

Table 7-1. The best scheme is the first, except perhaps for bottoms composition control in exceptionally large columns

Scheme No.	Manipulated Variable	Relative Advantages	Relative Disadvantages
1	Distillate flow	Internal reflux (decouple from energy balance)	Slower control of bottoms composition (stripping columns)
2	Reflux flow	Faster (independent of level loop)	Moderate interaction with energy balance
3	Reboiler steam	Fastest response (least dead time)	Severe interaction with energy balance
4	Bottoms flow	Slight interaction and fast bottoms composition control	Inverse response in sump level and separation

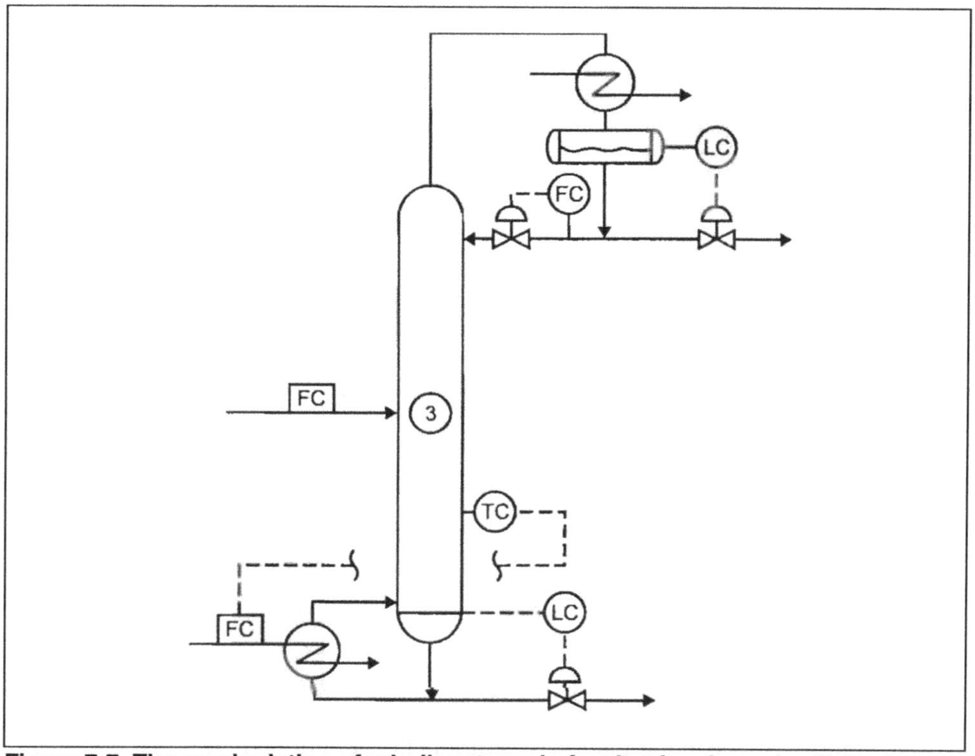

Figure 7-7. The manipulation of reboiler steam is fast but is adversely affected by a low gain in the bottoms level controller, thermowell lag, and severe interaction with the energy balance [3]

The manipulation of steam flow (separation) creates the greatest interaction between the material and energy balance and suffers the most from an excessive thermowell lag.

Exercises

7-1. What is usually the best column control scheme and why?

7-2. When is the manipulation of the distillate flow by temperature control not appropriate?

7-3. What is the single most effective advanced regulatory control strategy in improving column composition control?

7-4. What tuning method would be best for loops that control the column split?

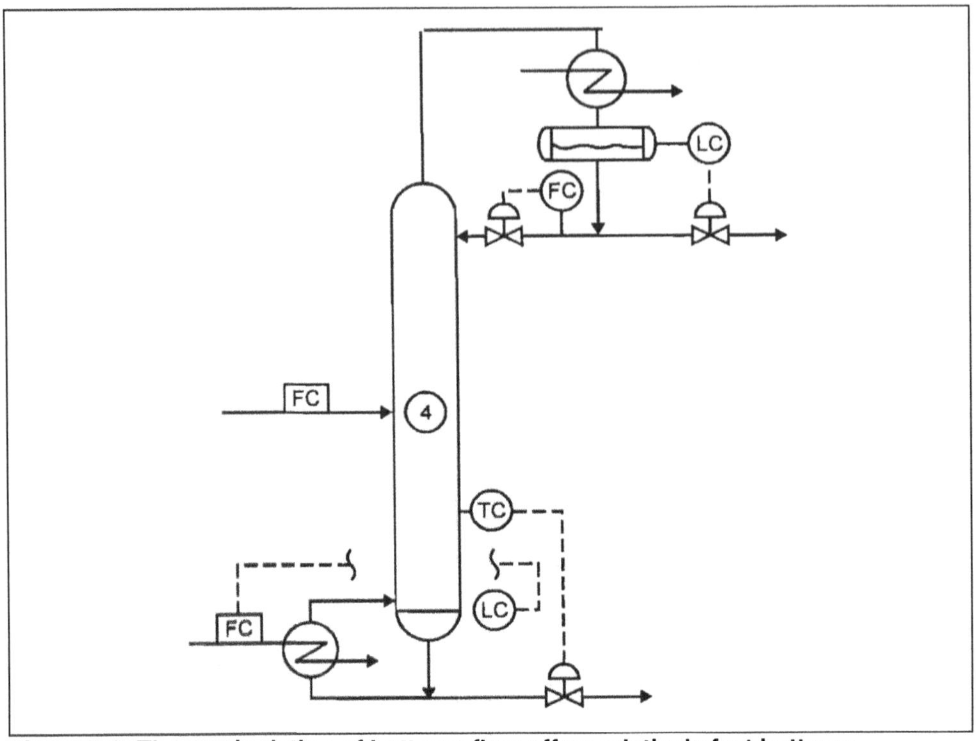

Figure 7-8. The manipulation of bottoms flow offers relatively fast bottoms composition control, but the inverse response in the sump can severely disrupt or disable level and separation control [3]

References

1. McMillan, G. K., *Tuning and Control Loop Performance,* 3rd edition, ISA, 1994.
2. Tolliver, T. L., "Improved Column Control to Reduce Distillation Operating Cost," *Chemical Engineering,* November 24, 1986.
3. Tolliver, T. L., "Distillation Control Design Based on Steady-State Simulation," *ISA Transactions,* Vol. 17, No. 3.

8

Vessels, Desuperheaters, Dryers, Kilns, Calciners, Crystallizers, Extruders, Chambers, and Rooms

This chapter describes control strategies and tuning requirements for a diverse mixture of process equipment and applications.

Learning Objectives

A. **Understand how the volume size and backmixing attributes of process equipment affect their controllability.**

B. **Develop the ability to minimize the interactions of temperature control loops.**

C. **Appreciate the implications of speed and tightness of temperature control for assorted applications.**

D. **Evaluate the benefits of alternative control strategies.**

8-1. Process and Equipment Design Considerations

For the purpose of discussion in this section, the vessel category is defined as storage, recycle, blend, or surge volumes where there is no heat of reaction or no agitator. Neutralizers fall into the reactor category because there is a heat of neutralization and a high degree of mixing. Many of the considerations for temperature control in vessels can be obtained from the discussion on reactor temperature control and from the equations in Appendix F by the omission of the heat evolution term and by attention to the much larger dead time due to slow dispersion and measurement lag times from poor mixing.

Fortunately, most vessel volumes are so large that the process time constant is extremely large and, consequently, the rate of change of temperature is exceptionally slow. This is fortunate because the only mixing typically provided in storage and feed tanks is provided by incoming flows (e.g., feed and recirculation) and convective currents. The addition of eductors and the location of dip tubes or spargers can greatly reduce the magnitude of the loop dead time. Ideally, thermowells should be installed in the elbows of recirculation lines. If they must be installed in the vessel, they should be located in areas of where the fluid velocity is largest, to reduce the measurement time constant, but farthest away from steam spargers, to avoid erratic nonrepresentative temperature readings.

Storage and feed tank temperature loops have a process time constant so large that a high controller gain can be used despite a high process dead time from poor mixing.

For desuperheaters, it is important that there be an adequate steam velocity in the desuperheater and a straight pipe run of 10 to 25 pipe diameters downstream for uniform water dispersion and cooling. Elbows act as great water separators and can cause water droplets that do not evaporate before reaching the temperature sensor. The result is an erratic temperature reading due to water impingement. Given the proper steam velocity and sufficient straight run, a distance of 20 to 45 feet from the injection to the measurement point ensures complete evaporation at the thermowell for set points that are at least 10 to 15°F above saturation.

Since feed irregularity from screws is the fastest and most common disturbance to dryers, kilns, and calciners, special attention should be given to the feed system to ensure a uniform load. Often there is a transition from batch centrifuge operation to continuous solids drying, or reaction or voids caused by poor distribution. The addition of a surge volume and error square inventory control by the adjustment of an amp controller set point for variable speed drives can help provide a constant load for processing. Screw live amps (i.e., total amps minus no load amps) act as an inferential measurement of product quantity and moisture content.

For kilns and calciners, sighting ports should be located for a clear, unobstructed view of product by optical pyrometers or optical fiber thermometers. The seals on rotating parts or at feeder entry and exit points should be as tight as possible to reduce heat loss and/or air infiltration.

Crystallizers can enter into regions of unstable operation or excessive nucleation. Variable speed agitators should be used to allow the optimization of mixing. Too little mixing increases the areas of supersaturation and nucleation rate that exceed growth rates. Too much mixing causes excessive particle attrition and secondary nucleation. Excessively cool heat transfer surfaces can cause localized nucleation. The result, in any case, is incorrect particle size distribution and the coating of heat transfer surfaces.

The buildup of crystals in passageways and entrances to circulation legs or pipe lines can cause excessive pressure drops and poor flow distribution that eventually lead to shutdowns for defrosting or cleanout. High flow and heated or slick surfaces (e.g., Teflon™ or polished), can minimize these interruptions.

Since premix heat and viscous heat have the biggest effect on extruder temperature and product quality, good mixing in the feed bin and variable speed drives are essential. Special connections should be designed for fast, nonintrusive temperature sensors.

The air flow rate, distribution points, measurement locations, and obstructions to flow patterns should be designed for maximum uniformity and minimum loop dead time for chamber or room temperature control. Sensor connections for inner (slave) temperature loops in large distribution ducts should be added.

8-2. Disturbances and Difficulties

The time constant to dead time ratios and, consequently, the peak error for temperature control are generally not as good for the process equipment in this section when compared with reactors or columns. The more longitudinal the equipment, the greater the tendency for plug flow to exist and most of the residence time to show up as detrimental loop dead time rather than as a beneficial process time constant. The exception is flash or fluidized bed dryers in which there is significant backmixing from turbulence.

There is significant interaction between the inlet and outlet temperature loops on dryers, kilns, and calciners, between zone temperature loops on extruders, and between the temperature and relative humidity loops on ducts for chambers and rooms. Classic decouplers can be designed, but these assume linearity and equal dynamics. They need to be adjusted for changing equipment and instrument and operating conditions. Alternative strategies and measurements discussed in Section 8-6 offer inherent decoupling and a more reliable solution.

The poor time constant to dead time ratio and loop interaction means that feedforward control should be considered to improve the control of dryers, kilns, calciners, and extruders. However, the lack of gravimetric feeders may render feedforward signals too inaccurate for effectiveness. Amp or power measurements can be used, but sometimes the difference between load and no load is too small to be a good indicator of feed. Flowmeters that use the impact of solids on a plate do not work well for wet material or a vibrating environment. The combination of speed and nuclear or microwave measurements are a possibility for some installations. Rotary valve speed offers the simplest solution if the entrance to the valve is full and the discharge is free.

Feedforward control is important improvement for dealing with the poor dead time to time constant ratio and interaction in dryers, kilns, calciners, and extruders.

8-3. Effect of Loop Performance

High or low temperatures in vessels may cause product degradation, undesirable reactions, and/or change of phase and an associated loss of yield, grade, or run time. In some applications, high temperatures may cause a safety issue due to exothermic polymerizaton. Poor desuperheater temperature control can damage downstream equipment.

Excessive drying can reduce the amount of water sold as product and cause hot or burnt spots. Insufficient drying may cause flowability or handling problems and result in recycled product or customer dissatisfaction. Poor inlet and outlet temperatures result in improper reaction temperatures and locations, respectively, in kilns and calciners, which translates to a yield or capacity of loss due to lower quality and conversion. Poor extruder temperature control often show sup as poor product quality. Poor temperature control in plant growth chambers can cause inconsistent research data, wasted time, and improper conclusions. Improper temperatures in sterilization chambers or spinning rooms can cause beverage and fiber quality problems, respectively.

Of course, excess heat supply or removal wastes energy. Dryers, kilns, and calciners may not be as big users of energy as columns but are generally larger than reactors or heat exchangers. Energy usage is naturally a bigger opportunity when electrical energy is used or when an additional boiler is forced online and its excess steam generation capacity at minimum fire is vented.

8-4. Controller Tuning

Due to the poorer time constant to dead time ratio, the process equipment in this section tends to have lower gains than columns or reactors. Gains typically range from one to four. Reset action for small volumes (e.g., desuperheaters) may be as large as 0.5 repeat/minute and for large volumes (e.g., storage tanks) could be as small as 0.01 repeat/minute. Rate settings are generally determined by the time constant of the temperature measurement and fall between 0.5 and 1.0 minute for reasonable thermowell designs and installations. Optical pyrometers and exposed optical fiber thermometers have a much faster response and necessitate a reduction in rate setting to 0.1 minute or less unless there are large volumes or heat transfer surfaces.

Large dead times, near dead time dominance, and interaction makes the temperature loops especially difficult to tune. Auto tuners are a useful tool except for cases of severe interaction. Slight interaction can be handled by detuning the least important loop.

8-5. Control Strategies

For throttling of coolant flow to coils or jackets of small vessels with a continuous discharge flow, the same concerns described for heat exchangers about a limit cycle from excessive dead time and process gain at low flow exist. Therefore, gain scheduling or signal characterization are useful techniques. Also, a feedforward signal can be computed, based on a simple energy balance as was done for the heat exchanger. For a zero discharge flow, set point profiles and proportional-plus-derivative (PD) controllers (like those used on batch reactors) can be used to prevent overshoot.

For desuperheaters, the water flow should be ratioed to the steam flow (i.e., flow feedforward) with the ratio corrected by the output of the temperature controller. If flow measurements are not available, inferential flow measurements from valve positions can be created for ratio control. A pseudo water flow controller can be created, as shown in Figure 8-01, by the addition of a filter of 0.05 minute to the calculated water flow to simulate the stroke and flow dynamics and the sue of a kicker on the signal to the control valve to compensate for valve dead band. The kicker simply adds a bias for an increase in controller output. Kickers can be applied to loops with actual feedback measurements to reduce the dead time from dead band, but a noise band and/or measurement filter must be added to prevent valve dither and self-inflected disturbances.

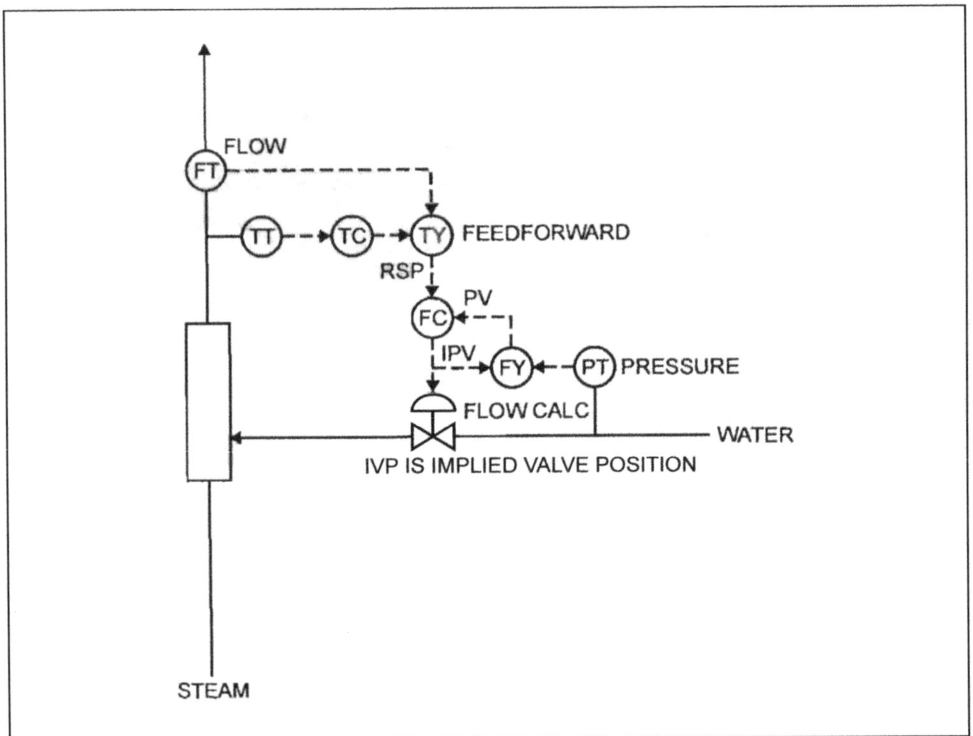

Figure 8-1. An inner pseudo flow loop for desuperheater control can be created to ensure a linear flow feedforward signal by use of a flow model and dead band compensation

The basic loop arrangement for dryer temperature control would use cascade control of outlet temperature to set inlet temperature. However, a constant outlet temperature does not mean a constant moisture content of product exiting the dryer.

The product temperature in dryers closely approximates the web bulb temperature as long as there is still some moisture present. Air flow through the dryer gives up sensible heat for the heat of vaporization. As a result, the movement of air through the dryer follows a line of constant dry bulb temperature and enthalpy on the psychrometric chart in Figure 8-2 [1]. The increase in dry bulb from inlet to outlet represents the moisture load. Of course, the minimum outlet temperature and, hence, the maximum moisture load is reached at the dew point temperature. If air flow can be manipulated directly or ratioed to feed rate, the dry bulb temperature difference between the inlet and outlet of the dryer can be used as shown in Figure 8-3 for moisture control of kilns and calciners [2]. This strategy to control moisture load helps eliminate the interaction from separate temperature loops and cancels out noise that appears on both temperature signals [3].

Below the critical moisture point, where inner particle moisture has been depleted and dry spots start to appear on the surface, the rate of drying is proportional to the moisture content. For this falling rate region, Equation 8-1 can be developed to equate moisture content to the driving force for moisture removal that is the log mean temperature between the air dry bulb and product wet bulb temperatures multiplied by some parameters. If the wet bulb product temperature can be measured or computed from an inlet air humidity or dew point and dry bulb temperature measurement, an inferential measurement of moisture can be directly computed via this equation. If this is not possible, an outlet temperature set point can be computed to provide inferential moisture control per Equation 8-2. However, this calculation creates positive feedback and will cause instability unless the coefficient R is set less than open-loop gain of the outlet temperature loop, and the time constant of the filter applied to the inlet temperature is set larger than the integral time (minutes/repe3at) of the outlet temperature loop [4].

Wet bulb temperature can be used to provide an inferential measurement or an outlet temperature set point can be computed from a filtered inlet temperature to provide dryer moisture control.

$$X = [(X_c * G * C)/(H_v * K * A)] * \text{Ln}\,[(T_i - T_w)/(T_o - T_w)] \qquad (8\text{-}1)$$

$$T_o^* = R * \text{fil}(T_i) + B \qquad (8\text{-}2)$$

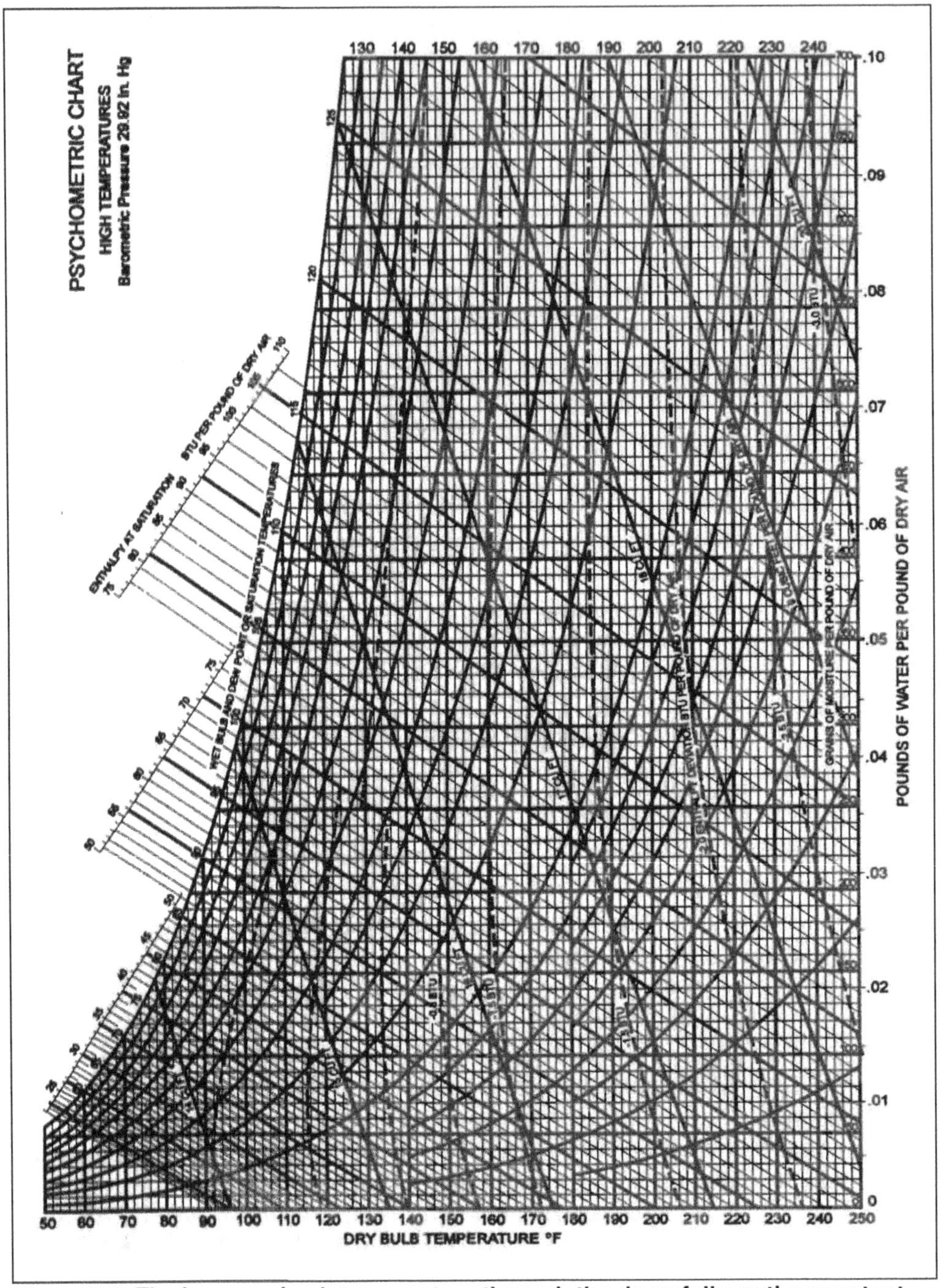

Figure 8-2. The increase in air temperature through the dryer follows the constant wet bulb temperature line on a psychrometric chart and indicates the moisture load [1]

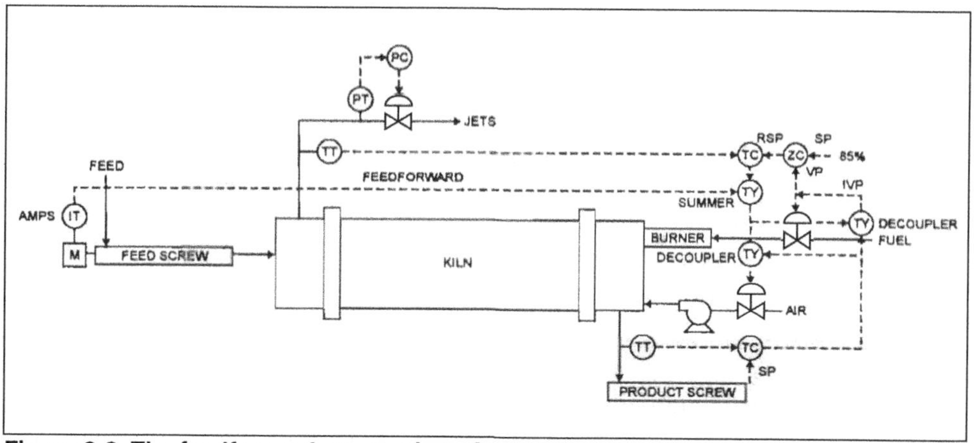

Figure 8-3. The feedforward summation of screw live amps and the other loop output reduces the interaction from upsets and the valve position controller maximizes throughput

where:

A = particle surface area (ft.2)

C = heat capacity (Btu/lb * °F)

B = bias (°F)

R = ratio of outlet to inlet temperature (dimensionless)

fil = filter

G = air flow (ft.3/min)

H_v = heat of vaporization (Btu/lb)

K = mass transfer coefficient

X = computed moisture (fractional)

X_c = critical moisture (fractional)

T_i = inlet temperature measurement (°F)

T_o = outlet temperature measurement (°F)

T_o^* = outlet temperature set point (°F)

T_w = wet bulb temperature measurement or calculation (°F)

The dryer temperature or inferential moisture control system can manipulate feed rate instead of heat input to maximize throughput. However, product voids or plugging will cause a high temperature that will increase feeder speed. A high temperature override controller needs to be added to cut back on heat input. Also, another mode of inventory control in the upstream bin must be used. If there is sufficient surge volume, the inventory can be loosely controlled by varying heat input. Otherwise, upstream production rates and centrifuge feed rate would have to be ratioed to dryer feed rate. An alternative method to maximize feed protection, without the concerns about high dryer temperature or upsets to upstream equipment, is to use a valve position controller to maximize a dryer fuel or steam valve by gradually increasing feeder speed. Upstream production rates would be slowly increased in concert.

In kilns and calciners, there are typically four zones. First is a drying zone of constant temperature, followed by a heating zone of rising temperature and a reaction zone of constant temperature, and finally a burning zone characterized by rapid rise and slow fall of temperature. The energy supplied in the drying and reaction zones are used to evaporate moisture and supply the heat to drive the endothermic reaction, respectively.

The basic loop arrangement for kilns and calciners would use (1) outlet temperature to adjust the heat pulled (e.g., air flow) through the drum and (2) inlet temperature to adjust the heat input (e.g., fuel flow). Note that the outlet air temperature loop has reverse action where an increase in temperature will cause a decrease in air flow. the outlet loop is not tuned as aggressively as the inlet loop. At steady-state or for small upsets, the interaction is acceptable. However, large upsets require the use of feedforward or decouplers to prevent excessive cycling. Throughput can be maximized, as shown in Figure 8-2, by the addition of valve position controller to gradually trim the outlet temperature set point and the reaction point along the drum to keep the fuel valve at its largest controllable position.

At absolute gas temperatures above approximately 1100 kelvins, the heat transfer is almost entirely by radiation, and the gas temperature is subsequently determined by excess combustion air. Adjusting the fuel or excess air will affect the excess air and change the heat transfer to the reaction and burning zones. The convective heat transfer to the cooler zones then depends upon the amount of heat left after radiation to the hotter zones.

An innovative control strategy illustrated in Figure 8-5 has been implemented to stabilize lime kiln control by the use of a fast oxygen loop to control excess air by manipulation of the forced air damper. A zirconium oxide probe is used to ensure the fast, reliable response needed for cascade control. The hot end temperature loop manipulates the oxygen set point,

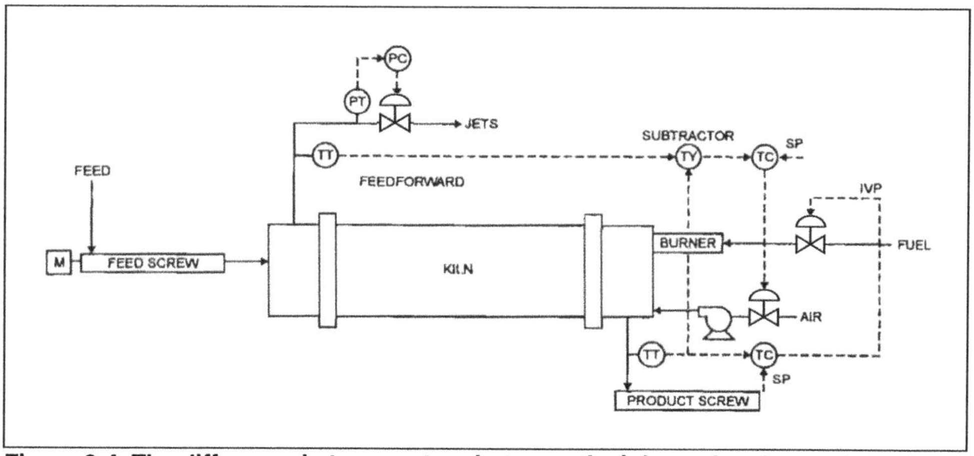

Figure 8-4. The difference in temperature between the inlet and outlet can be used to control moisture by manipulation of air flow

and the cold end temperature loop manipulates the fuel input. A feedforward signal from the fuel valve position that represents the air-to-fuel ratio is used to assist the oxygen loop. The feedforward signal must take into account the installed characteristics of both the fuel valve and the air damper. The kiln pressure set point is trimmed based on fuel input to set the flame length [5]. This control strategy implemented with model predictive control has achieved significant energy savings and improvements in product quality.

An innovative kiln control strategy that uses oxygen, feedforward, and decoupling best implemented by model predictive control offers significantly tighter control and energy savings by dealing with interactions and difficult dynamics.

Batch-operated stirred cooling crystallizers are the most common type in the industry, but they tend to yield poor quality, primarily due to the high cooling rate early in the batch cycle. A large number of nuclei are formed that cannot grow to the desired size. Also, crystal growth (frosting) on the heat transfer surface can be intensive and cause insufficient cooling throughout the rest of the batch. These problems can be largely overcome by the use of the programmed cooling curve, as shown in Figure 8-6, that is the reverse of the natural cooling curve.

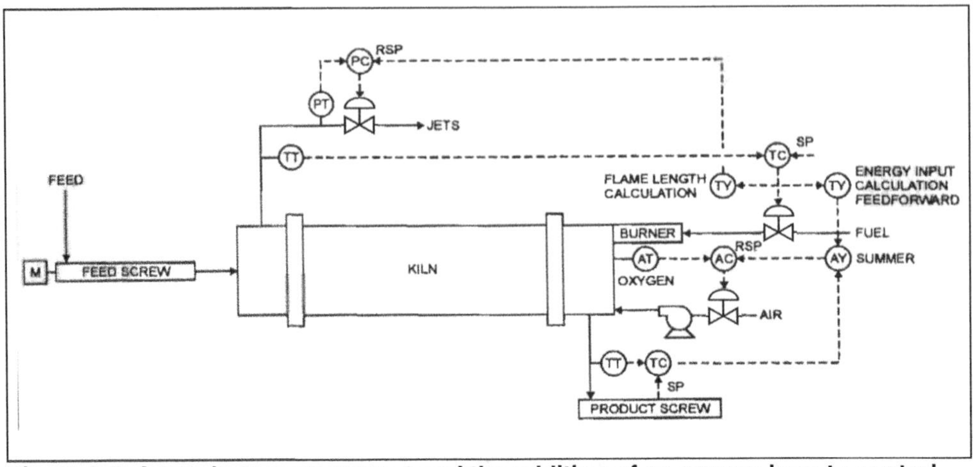

Figure 8-5. A new loop arrangement and the addition of an oxygen loop to control excess air leads to more stable kiln control [5]

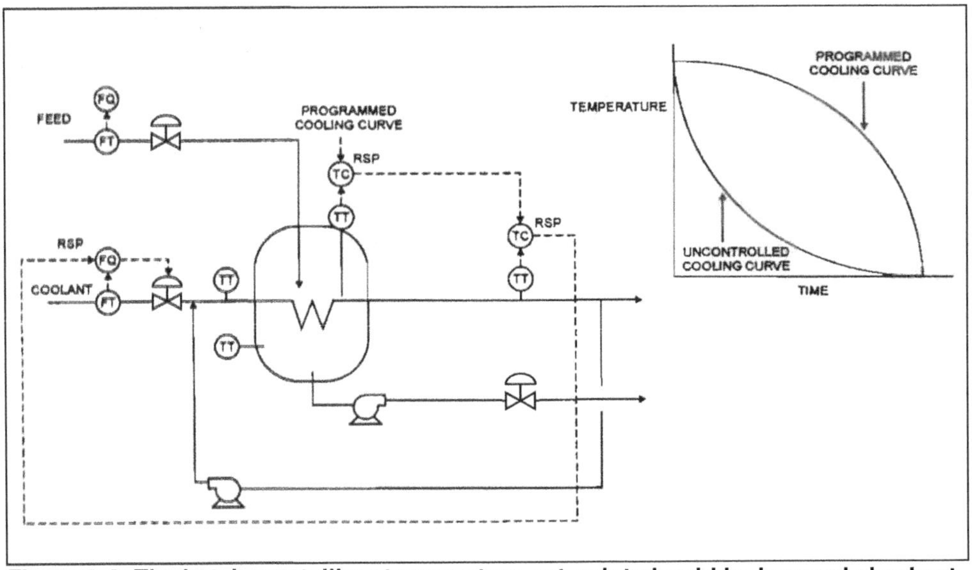

Figure 8-6. The batch crystallizer temperature set point should be lowered slowly at the start and rapidly at the end to provide a cooling curve that optimizes crystal growth

An optimum cooling curve starts with low cooling rates and finishes with maximum cooling rates prevents small crystals and frosting of heat transfer surfaces.

Figure 8-7 illustrates an elliptical optimum quality region on a plot of specific energy consumption (SEC) versus feed rate for extruders. The curves represent various amounts of heat added to the premix. During start-up, the goal should be to get to an operating point within the innermost ellipse as quickly as possible by the adjustment of energy input within the constraints of shear and die pressure. The SEC curve needs to be reproducible but does not necessarily have to match the manufacturer's curve. An irregular shape or noise indicates the efficiency factor of the screw or mixer used in the SEC calculation should be adjusted. Since the energy input from screw speed or premix heat is much larger than that from zone heaters (see the FORTRAN subroutine in Appendix G for extruder simulation for the relationships), the screw speed or premix heat should be manipulated to control SEC at the optimum. The zone heaters provide essentially just a warm blanket.

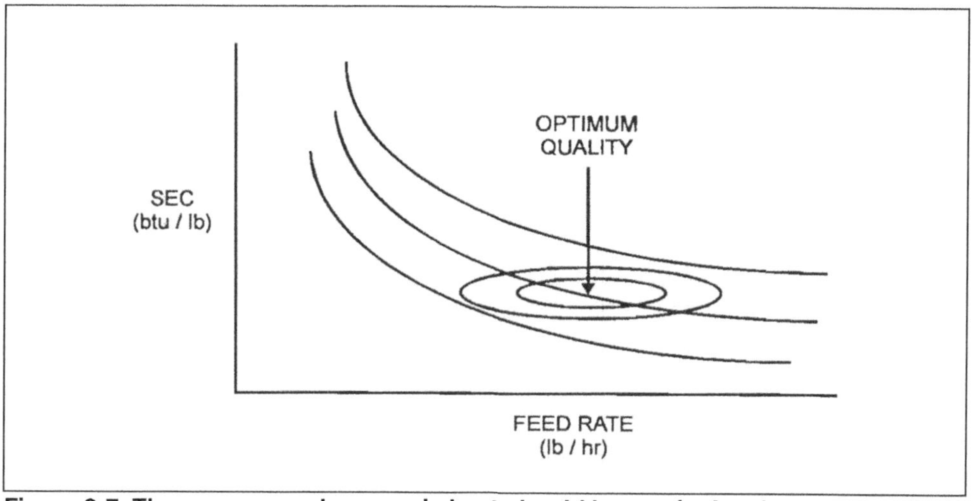

Figure 8-7. The screw speed or premix heat should be manipulated to control a computed specific energy consumption (SEC) at a set point in the innermost ellipse for optimum quality (Rx slide)

The temperature control of chambers and rooms benefits from the cascade control of room temperature to duct temperature. The inner (slave) duct temperature loop corrects for upsets from air sources and helps linearize the outer room temperature loop. The use of dew point temperature instead of the relative humidity loop eliminates the interaction with the dry bulb temperature loop. An enthalpy computation can be used to save energy by optimizing the split between return versus outside air for spinning rooms [6]. The major sources of dead time are (1) dead zones due to obstructions or poor air circulation patterns, (2) room air turnover time (i.e., room volume divided by blower capacity), (3) sensor lag caused by low velocity, and (4) damper dead band from slope in linkages.

Exercises

8-1. What are some of the difficulties associated with desuperheater control?

8-2. What is the single most likely problem with the implementation of dryer inferential moisture control?

8-3. What are typical problems encountered with kiln temperature control?

8-4. What is the worst possible control strategy for a crystallizer?

8-5. What are some possible improvements for room temperature and humidity control?

References

1. Shinskey, F. G., *Energy Conservation through Control*, Academic Press, 1978.

2. Robinson, J. W., "Improve Dryer Control," *Chemical Engineering Progress*, December 1992.

3. Robinson, J. W., "A Unique Drying Moisture Control System," *Sensors*, June 1992.

4. Shinskey, F. G., *Process Control Systems*, McGraw-Hill, 1988.

5. Blevins, Terry, and Rice, Rodney, "Automating a Lime Kiln Control," *Tappi Journal*, March 1983.

6. McMillan, G. K., "A Distributed Control System for Enthalpy Control of Air Washers," ISA, New Orleans, May 1987, Vol 19.

9
Wireless

Learning Objectives

A. Learn about the latest technology advances in wireless networks.

B. Understand the importance of key features, such as security, which are essential for every wireless sensor network.

C. Learn terminology and issues to intelligently discuss wireless industrial applications.

D. Understand the impact of enhanced PID technology when performing control across wireless sensor networks.

E. Recognize the advantages of wireless smart measurements.

F. Understand the importance of exception reporting when supporting loops with fast dynamics.

G. Learn the importance of mesh technology when considering installation requirements.

H. Understand how wireless mesh networks can be used in monitoring and control applications.

I. Understand how wireless mesh networks can be used to improve overall plant operations.

9-1. Introduction and Overview

Smart devices are an essential part of nearly every control system. These smart devices provide greater process insight, reduce engineering costs, and otherwise contribute to improving the overall operational performance of the plant. Smart devices offer advanced diagnostics that can report the health of the devices and in many cases, the health of the process that the devices are connected to. It is not uncommon for smart devices to include diagnostics that can detect plugged sensing lines, burner flame instability, loss of agitation, wet gas, orifice wear, leaks, or cavitation. These devices tell the user how well they are operating and when they need maintenance. Improvements in sensor technology and improved diagnostics have led to a large variety of smart instruments. Wireless technology can allow users to more readily migrate to smart instruments and valves by eliminating the need to add smart input and output (I/O) cards for the distributed control system (DCS) and reducing the time for commissioning. For a more fundamental understanding of sensor and transmitter roles and capability, see Chapter 1 of *Essentials of Modern Measurements and Final Elements*.

Smart devices incorporate features such as status on measured values, time stamps, event latching and confirmation, and block data transfer. The availability of additional variables, such as density and temperature for Coriolis mass flowmeters, provides additional information useful for understanding changes in the process. Using these smart devices results in fewer trips into the plant, a much better controlled process, and significantly less downtime.

The advantages of wireless networks go well beyond the obvious initial savings in the design, installation, and commissioning of new smart instrumentation. The bigger benefits are seen in the lifecycle cost of smart instrumentation and the more effective and expansive use of smart instrumentation capabilities. As discussed in Chapter 1 of *Essentials of Modern Measurements and Final Elements*, smart instrumentation installation errors and drift are dramatically reduced by advances in sensing element design and the compensation of installation conditions and sensor response. This reduction in onsite process calibrations requirements combined with faster troubleshooting by the use of online diagnostics and, with wireless networks, the elimination of wiring noise and connection problems translates to significant maintenance cost savings [2].

Unfortunately, utilizing the features of these smart devices is not always possible with the installed base of control systems. What is needed is an infrastructure that can be used with both new and existing system installations. Wireless sensor networks, in particular WirelessHART™, offer a low-cost, effective, and rapid deployment of smart devices in both new and existing installations.

WirelessHART (Highway Addressable Remote Transducer) was designed from the ground up to support the capabilities of new systems while at the same time not leaving behind the large installed base of wired HART devices. Wireless communication can be easily added to existing wired HART transmitters by the simple addition of a Thum™ communicator (a device that connects to and is powered by the existing wiring in the transmitter and communicates via an antenna, additional process variables and diagnostics not previously available due to a lack of intelligence in the DCS I/O). Additionally, WirelessHART transmitters can replace non-smart transmitters or provide new smart measurements. The replacement or addition of wiring and DCS I/O cards is not needed. WirelessHART supports existing applications such as pressure loops while at the same time supporting new applications such as vibration, monitoring and discrete applications. HART Discrete Field Device support is targeted at devices such as field powered electric actuators, discrete on-off motor starters, field switches for detection of abnormal operating conditions, proximity switches, photo sensors, contacts indicating equipment status, field pushbuttons, etc. Device manufactures will include HART diagnostics which in turn will differentiate HART Discrete Field Devices.

Wireless instruments enable a fast and low-cost installation for upgrade projects to take advantage of the advances in sensor technology and diagnostics.

Wireless instruments offer portability for optimizing measurement locations, analyzing profiles, and troubling shooting equipment by online performance metrics.

Good control performance requires that the overall control loop execute four to ten times faster than the process time constant plus deadtime. When calculating overall control performance, the amount of time required to read input signals, execute control, and drive outputs must be taken into consideration. These times are referred to as *latency*. Variation in input, output, and control execution times is referred to as *jitter*. Too much latency or jitter can severely degrade overall control performance [1][4]. Managing latency and jitter in the input and output operations in wireless networks is challenging. Wireless signal strength varies in time; wireless nodes can become unavailable; data may be routed dynamically on different paths. Oversampling burns batteries out prematurely. Control performance is degraded if process data is missing or delayed in the network [1][3].

Wireless sensor network technology can be viewed as a platform for new kinds of applications, as well as a platform for smart devices in existing plant environments [22]. WirelessHART targets unit-level process operations and supports the following applications:

- Equipment and process monitoring
- Control
- Asset management
- Diagnostics/predictive maintenance
- Location tracking
- Nomadic, "wireless worker" applications
- Portability for optimization of installation location
- Research and development in labs and pilot plants

Background

HART technology has been in existence since the late '80s. In its initial release the HART Field Communications Protocol was superimposed on a conventional 4- to 20-mA signal, providing two-way communications with smart field instruments without compromising the integrity of the measured data. During its 20+ years of existence, the HART protocol has evolved from a simple 4- to 20-mA-based protocol to the current wired and wireless-based technology with extensive features supporting security, unsolicited data transfers, event notifications, block mode transfers, and advanced diagnostics (Figure 9-1) [5]. As has been mentioned, diagnostics now include information about the device, the equipment the device is attached to, and in some cases the actual process being monitored.

Compared to wired HART, WirelessHART incorporates many new features providing improved performance, enhanced diagnostics, and better maintenance capabilities [5]. WirelessHART is a secure networking technology operating in the 2.4-GHz industrial, scientific and medical (ISM) radio band. It utilizes IEEE 802.15.4–compatible direct-sequence spread spectrum (DSSS) radios with channel hopping on a packet by packet basis [5, 6]. WirelessHART communication is arbitrated using the WirelessHART Network to schedule link activity. A given communication link may be dedicated to communication between a network pair or the link may be shared. WirelessHART networks are organized as a mesh

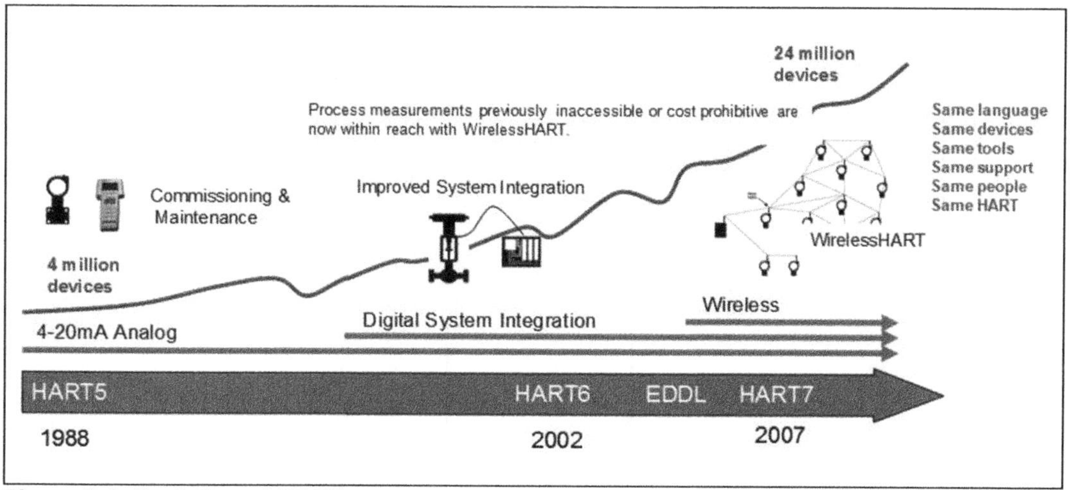

Figure 9-1. Evolution of HART

topology. In a mesh topology, each device is connected to several other devices and all devices are capable of routing communications traffic. Devices can be connected to each other through several hops. A wireless device network (Figure 9-2) supports a wide variety of devices from many manufacturers.

Mesh networks are self-healing: the network can still operate even when a device breaks down or a connection goes bad. As a result a mesh topology is very reliable.

Improved Measurement and Control

Wireless device networks are used in a wide variety of applications including monitoring and controlling tank levels, monitoring emission levels and water quality, monitoring equipment health, and a wide variety of control applications. Measurements are often used to generate reports for the FDA, EPA, and other agencies. These measurements are also used by plant personnel to make decisions about the operation of the process, plan maintenance activities, schedule production runs, and validate the quality of the finished product. In many systems, these wireless measurements are part of feedforward and feedback control strategies.

There are many kinds of final control elements including valves, agitators, blowers, conveyors, etc. These control elements receive a signal and perform some action. In the case of an on-off valve, the valve will attempt to open or close; valve status must then be communicated. In the case of a regulating valve, the read-back of the actual valve position is important in

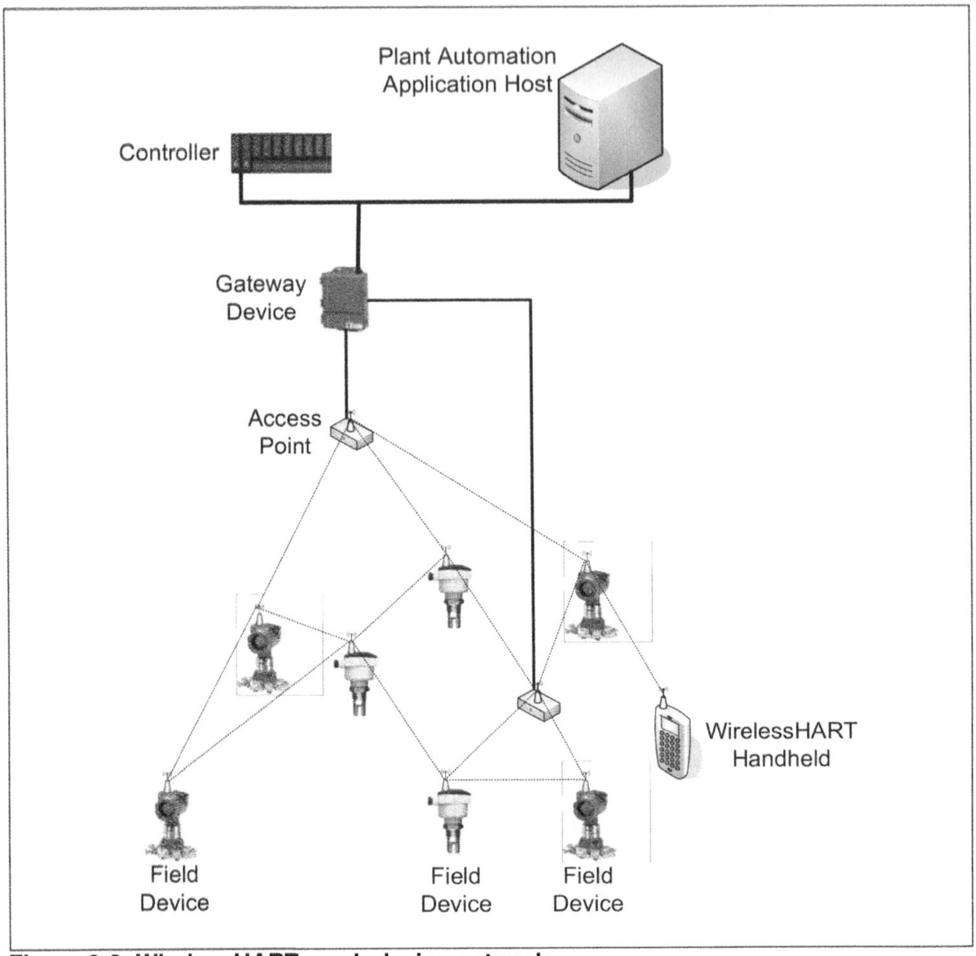

Figure 9-2. WirelessHART mesh device network

order to determine if there are problems with the final control element itself that is impacting control performance. In the case of an agitator, it's important to know if the agitator is moving and, perhaps, at what speed. Newer final control elements are engineered to include a radio as part of the device; older devices can be connected to a wireless network using an adapter [5].

An important factor for DCS and host applications is the device status that is communicated with measurements. Device status provides an indication of the measurement quality ("level of goodness"), whether the device making the measurement is healthy, and the timeliness of the measurement. Devices perform checks on hardware and software associated with the input or output and, in turn, report this through the status value.

The status of an output parameter is calculated to specifically indicate the quality of the value: good, poor, bad, or fixed. A "good" signal may be used for control. A "poor" value is suspect and may not reflect the true measurement or calculated value. A "bad" value means that the parameter value does not reflect the true measurement, calculation, or control value. "Fixed" means that the parameter value is constant and is not being updated in a periodic fashion. When a status other than "good" is received, the control system must reflect the status value in its diagnostics, and in the case where the value is used in control, the control strategy must either take the status into account or switch the control loop to manual.

Status also reports additional information such as configuration changed, cold start, loop current fixed (in wired systems), loop current saturated, non-primary variable out of limits, and primary variable out of limits.

All WirelessHART devices also include an Extended Device Status capability. The Extended Device Status value provides information on whether the device has malfunctioned (Maintenance Required), a device variable (e.g., temperature or pressure variable) is in an alarm or warning state (Device Variable Alert), or when power is critically low (Critical Power Failure). Both device status and Extended Device Status are included as part of burst mode communications. Burst mode communications is featured whereby devices send digital measurements and computed values on a periodic or exception basis. Burst mode with exception mode turned on results in a significant reduction in communications.

Another feature of WirelessHART is the time stamp that is communicated with all measurement values. The addition of the time stamp allows applications to determine how current the value is and whether or not there is jitter in the measurement, and in some cases, to use the time value to determine appropriate control actions using the measurement value.

Many wireless devices, such as analyzers, vibration monitors, and more complex devices, generate multiple digital measurement values. WirelessHART allows each burst communication to communicate up to eight digital values. If more than eight digital values need to be communicated, multiple burst communications can be sent.

Improved Operations

Operators provide a key function in plant operations. How well the plant operates is directly correlated to how well an operator is able to assess the state of the process (i.e., is the plant operating at its design set point), interact with equipment, respond to process conditions, and communicate with other plant personnel. Operators are responsible for starting up and shutting down equipment, adjusting the equipment to meet production schedules, and responding to unplanned situations, such as equipment failures, power changes, etc. When things aren't working correctly, they are the first line of defense.

Consider the following scenario in which the temperature of the process fluid in a tank is controlled by a valve and the temperature measurement is oscillating about its set point. As a first step in investigating this problem, the operator would likely put the valve position target, the actual valve position, the temperature reading, and the temperature set point on a trend. After reviewing the trended oscillations, the operator may decide to adjust the controller gain. In this scenario, the operator notes that the oscillations remain the same regardless of whether there has been an increase or decrease in controller gain. The operator also notes that the read-back value of actual valve position indicates that the minimum change in valve position is 0.5% and notes that a step in the actual valve position change always precedes the temperature change. The operator suspects that the oscillation is being caused by the resolution (stick-slip) of the control valve. Following through on this the operator requests a valve signature from the valve and compares what he or she is seeing with the actual process.

In order to support this scenario, several key functions must be supported. First, the sampling and reporting must be fast enough to not to miss the oscillations. Second, the actual valve position and signal status must be reported along with the target position. Third, the control valve needs a standard way to report more complex information such as the valve signature.

WirelessHART has extensive support for all of these requirements. In this case, the operator would set the sampling interval to a fast rate and enable exception reporting. With exception reporting, even though devices are configured to report at a fast rate, they only communicate if the measured value deviates by more than a set amount. To support the second requirement, multiple values can be communicated in the same packet. To ensure that both upstream and downstream communications get through, graph routing is used in both directions (graph routing is a set of directed paths that connects one device to another device). The third requirement, accessing the valve signature, is addressed using Block Data Transfer or Trending. Trending is intended to allow devices to collect data at faster more precise intervals while at the same time reducing the number of communications required to transport the data from the device to the host. Block Data Transfer is used to create a connection between the DCS or host and the remote device. Block data transfer is designed to maximize the utilization of the communication bandwidth while performing the transfer.

Operators are responsible for a large number of control loops, many pieces of equipment, and in many cases, more than one process or utility area. To support this large scope, operators rely on a well engineered alarm system. Since smart devices analyze themselves and report their health on an ongoing basis, they are ideal components of an alarm system. As we have seen, smart devices also have the ability to provide additional information about the process they are inserted into—for example, they can detect plugged sensing lines, burner flame instability, loss of agitation, wet gas, orifice wear, leaks, bubbles in the line, and cavitations. Both wired HART and WirelessHART support these capabilities through event notifications. In the case of WirelessHART, communication priorities and buffers have been optimized to support event notification.

"Event notifications" include the device's status byte, Extended Device Status byte, and the device's device-specific status information. It is possible to specify a limited set of bits that will trigger event notification. To prevent spurious event notifications a De-bounce Interval is configured. This defines the amount of time that a condition must persist before the event notification is time stamped and sent out. Once an event has been latched (once a condition has been detected the corresponding bit is set and remains set until it is acknowledged), it is transmitted repeatedly at the rate indicated by the Retry Period until the event has been acknowledged. Event notifications include a time stamp representing the first time a notification occurred.

Configuration, Installation, and Checkout

WirelessHART builds on HART. The HART infrastructure provides an extensive set of features supporting configuration, calibration, installation, and checkout. Devices are described in Electronic Device Description Language (EDDL). EDDL describes the devices' resource (hardware, electronics, and diagnostics information) and transducer information, methods, and user interfaces. Since new devices and new types of measurements will be added over time it is important that WirelessHART maintains both forward and backward compatibility. WirelessHART configuration conforms to the IEC 61804-3 standard [23].

As part of the installation process, field devices are configured and calibrated for the specific process where they are being used. As part of the configuration, each device is given a tag and scaled (device scaling includes high scale value, low scale value, engineering units, and decimal places), and signal conditioning is applied (e.g., in the case of a valve, it is important to know which direction the valve moves when the signal value is increased). A field device may be commissioned and calibrated at the factory, on site in the instrumentation shop, or after the device has been installed.

The set of parameters a device supports and the methods available to the device are described by the device's description. Users usually have standard parameter value templates for different types of devices and use these templates to customize devices for use in their plants. Tools such as the AMS 375 field communicator can be used to configure and calibrate field devices.

Device measurements are communicated to control systems or host applications. The parameters to be communicated and the rate that they are communicated depend on the process that the device is being used in. Configuring control systems and host applications requires a different set of information. In the case of the control system what is important are measurements that the device supports and the signal information that is associated with those measurements. Signal information includes information such as signal tag (since devices can be multivariable they can have many signal tags associated with them), scaling information, alarm information, linearization, and description. Control system configuration is performed off-line and downloaded into the control system at various points in the configuration cycle.

9-2. Principles

Supporting process automation applications requires consideration of each of the following: latency, energy use, robustness, security, and performance. Managing latency, while at the same time conserving power, is a difficult problem. To minimize latency, communications need to be organized so that packets are not delayed en route from source to destination. To minimize energy use, devices should be kept in a low-power mode as much as possible [7][13]. Deployments must be easy to set up, and once deployed, devices must run unattended for long periods of time, usually years.

At the communication distances typical in wireless sensor networks, listening for information on a radio channel costs about the same as data transmission. Furthermore, the energy cost for a device in idle mode is approximately the same as for a device that is in receive mode. Thus, the biggest single action to save power is to turn the radio and the device itself off during idle times, i.e., put the device to sleep. Turning the device off implies advanced knowledge about when the device will be idle. The approach taken by WirelessHART is to configure the device with knowledge about when it should wake up, perform some function, and go back to sleep; this configuration is performed by a network manager and is called scheduling [5][8].

In WirelessHART, communications are precisely scheduled using an approach referred to as Time Division Multiple Access (TDMA). The vast majority of communications are directed along graph routes. As mentioned, scheduling is performed by a centralized network manager that uses overall network routing information in combination with communication requirements that devices and software applications have provided. The schedule is subdivided into links and transferred from the network manager to individual devices; devices are only provided with the links where there is a transmit or receive requirement. The network manager continuously adapts the overall network graph and network schedule to adjust for changes in network topology and communication demands.

Several key features of the WirelessHART Network, such as self-healing, self-organization, and redundant routing, combine to make the wireless network robust. The network manager is responsible for the creation of this schedule and the associated connections. It is also responsible for the distribution of this schedule to the individual network devices. This scheduling function may be broken into the following phases [5]:

1. Support devices joining the network. As part of this the network manager is responsible for authenticating and orchestrating the join process.

2. Establish routes. As part of this the network manager is responsible for the creation of routes that can be used by plant automation hosts, gateways, other devices, and the network manager itself to perform communications with the application layer in network devices (the application layer in a device is everything above the network communication stack, for example in pressure transmitter this pressure measurement, device diagnostics, etc.).

3. Schedule communications. As part of this the network manager is responsible for the establishment of superframes and links that the user layer application of a network device may use to transfer process data, alerts, diagnostics, and other traffic to a gateway device for access by the plant automation host. The superframes also include links for network management and the join process.

4. Schedule control functions. For network devices that are actuators, interlocks, or any device that affects the process, the network manager is responsible for the establishments of routes, superframes, and links that the plant automation host may use to send set points and outputs to the user layer application in field devices.

5. Adapt the network. The network manager continually collects data from devices on the health of connections and traffic patterns and uses this information to adjust routing and scheduling.

The effectiveness of the overall network ultimately boils down to a combination of routing and scheduling. The network manager develops a set of routes and then uses these routes in combination with requests for network resources to schedule the network. The schedule is then distributed to each device using network management commands. Each device contains device management, which is responsible for communicating with the network manager and storing network configuration in the device. The overall communication stack (Figure 9-3) follows the International

Organization for Standardization (ISO) Open System Interconnection Reference Model (OSI Reference Model or OSI Model) [12].

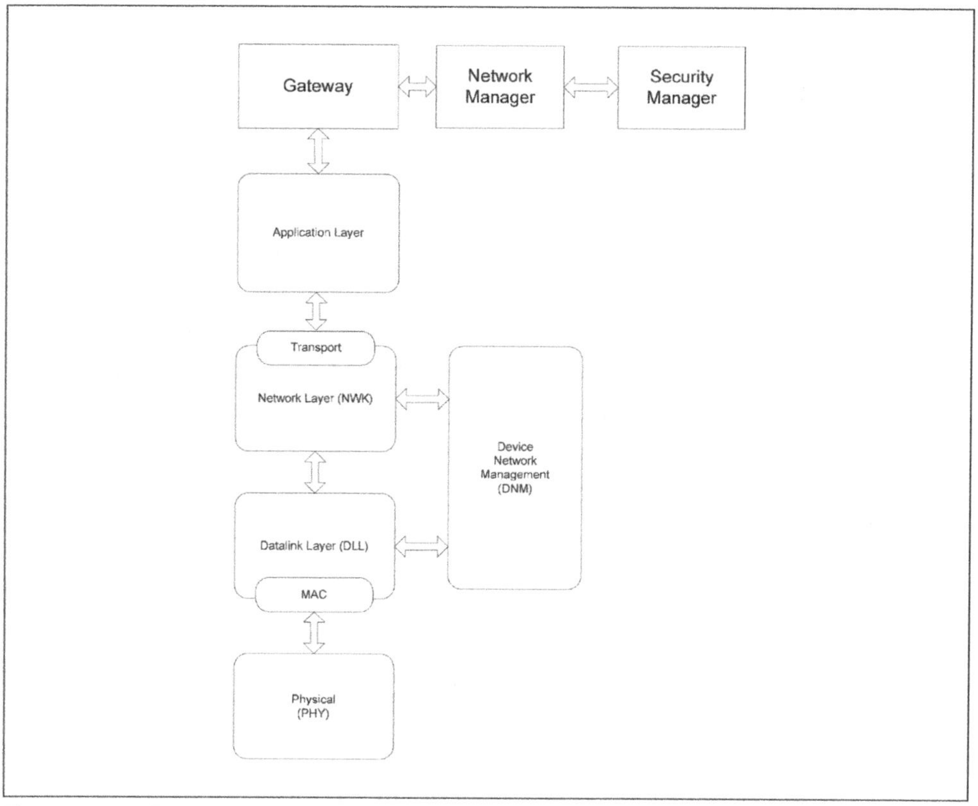

Figure 9-3. WirelessHART communication stack

The wireless network achieves very high reliability through the use of several mechanisms: multiple paths to devices, multiple RF channels, and multiple communication tries. If improved reliability is required, more paths can be inserted by the network manager. As additional devices are added to the mesh network the strength of the overall mesh gets better and better. Additional devices improve path diversity. Additional access points, and devices in general, increase throughput, reduce latency, and can be used to route around areas of potential interference [7, 8].

All communication on the wireless sensor network is time-synchronized. The basic unit of measure is a time slot, which is a unit of time of fixed (10 ms) duration commonly shared by all devices in the network. The duration of a time slot is sufficient to send or receive one packet and an accompanying acknowledgement, including guard-band times for network-wide synchronization (a guard band is a frequency that insulates one signal from another).

At the lowest level, communications are divided into time slots. Since WirelessHART supports up to 15 channels, 15 time slots may be in operation at the same time. To improve reliability, the physical channel is alternated at run time using an algorithm described in the WirelessHART standard. Within each time slot, devices are scheduled to transmit or receive. If there is only one transmitting and one receiving device the time slot is designated as a dedicated slot (a slot is also referred to as a link—the term *link* is used to define the association between devices during a specific time slot). If more than one transmitter can transmit into the link, the link is designated as shared. In this shared case, the transmitters must perform a clear channel assessment (CCA) before they attempt transmission. Should simultaneous transmissions (collisions) occur, devices use a backoff algorithm to manage communications. If more than one receiver is enabled to receive, the link is designated as a broadcast time link.

As has been mentioned, precise time synchronization is critical to the operation of the sensor network. Since all communication happens in time slots, devices must have the same notion of when each time slot begins and ends, with minimal variation. Time propagates outwards from the gateway (a gateway is a device that is used to connect wireless network control systems and higher level business applications) (Figure 9-4).

Communications are organized by superframes and time slots. As shown in Figure 9-5, in each time slot there must be one transmitter and one or more receivers (one for dedicated links; multiple for shared or broadcast). In this example, devices A and B communicate during link 0; devices B and C communicate during link 1; and link 2 is not being used. Every three links, the schedule repeats. An overall schedule may have many superframes. Real superframes would typically be hundreds of time slots long.

9-3. Security and Reliability

Communications across a wireless mesh sensor network must be both secure and reliable. Security is inherent in the design. To join a WirelessHART network, devices must pass a rigorous join process during which they authenticate themselves with the Network manager and in

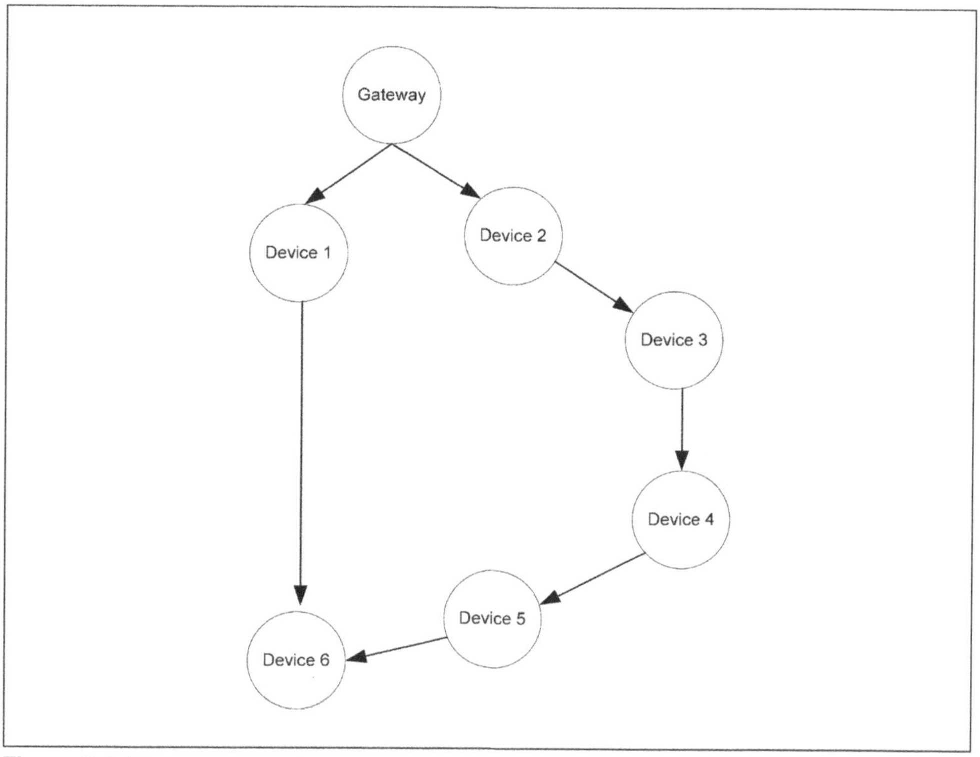

Figure 9-4. Time propagation

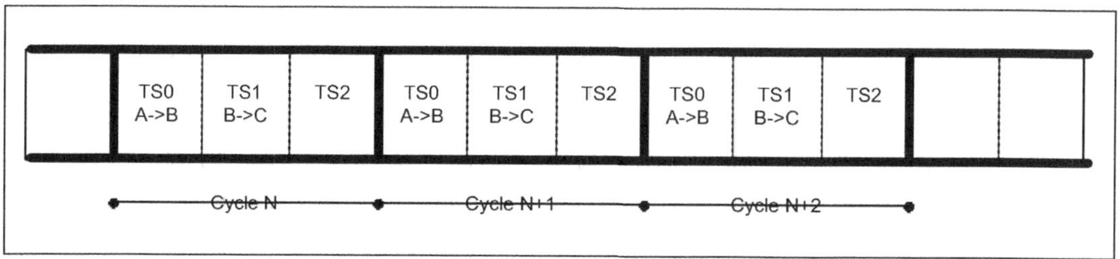

Figure 9-5. Example of a three-slot superframe

turn receive session keys [11]. All communications in WirelessHART are encrypted.

Communications are secured by ensuring that only devices approved to join the network join, that communications are not tampered with, and that bad guys cannot spoof the network through the use of replay attacks (a replay attack occurs when the same packets are resent on the same network). Communications are made reliable by retrying on different channels, retrying at different points in time, and communicating on different paths.

A security manager works with the network manager to secure the network from threats to its operation. The security manager generates and manages the cryptographic material used by the network. It is responsible for the generation, storage, and management of keys (Figure 9-3).

All communications in the network are encrypted and enciphered using the Advanced Encryption Standard (AES) [9, 10]. The standard comprises three block ciphers, AES-128, AES-192 and AES-256, adopted from a larger collection originally published as Rijndael. AES was announced by National Institute of Standards and Technology (NIST) as U.S. FIPS PUB 197 (FIPS 197) on November 26, 2001 [10]. A mesh network uses AES-128. The AES ciphers have been analyzed extensively and are now used worldwide. Many communication chips now contain a built-in AES coprocessor [20].

Joining the Network

As was mentioned above, devices that wish to join the sensor network must be authenticated through the join process. As part of the join process, devices are initialized with a join key and a network ID. Devices that are part of the existing network advertise for new neighbors; devices wishing to join the network listen for these advertisements. Devices wishing to join the network use their join key to encrypt a join request that they send to the network manager. The network manager uses the unique ID for the device to look up the device's join key, decrypts the join request, and if everything checks out, allows the device to join the network. The network manager then completes the join process by providing the joining device with a set of superframes, links, routes, and graphs. Join keys can be unique for each device or shared—the strategy is determined by the user and set in the gateway. Once a device joins the network the network manager can change the device's join key.

Securing Communications

All communications across the sensor network contain a security header (Figure 9-6); the security header provides a bridge between the security sub-layer in the source and the destination. The security header is designed to ensure that the payload is not tampered with. The security header starts with a security control byte that specifies the type of security that is being used. The counter is part of a nonce counter. Nonce counters are used to detect and handle reply attacks. The Message Integrity Code (MIC) ensures that the enciphered payload has not been tampered with.

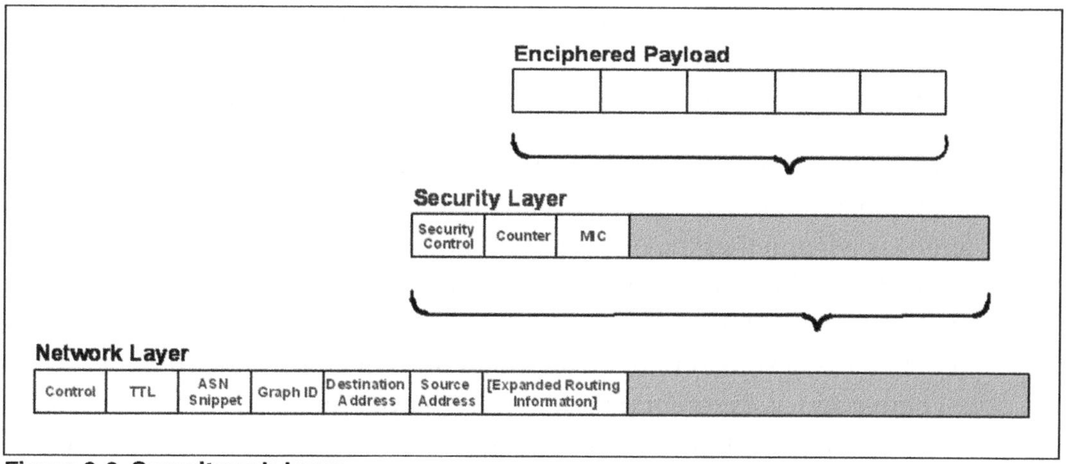

Figure 9-6. Security sub-layer

Sessions

Wireless sensor networks must seamlessly support services for a wide range of applications, including real-time data, alarms and events, block data transfers, configuration, calibration, and diagnostics. Sensor networks support communication traffic sourced by the network manager, by the gateway, by host applications, by devices, by location devices, and by users through host level applications and handhelds. To support these applications, session-oriented communications are used. Through these sessions applications have access to network resources (network resources are the channels and bandwidth provided by the network). Each device supports the following four sessions:

- Unicast network manager communications
- Broadcast network manager communications
- Unicast gateway communications
- Broadcast gateway communications

Additional sessions must be added for handhelds and peer-to-peer communications. Only the network manager can add or remove sessions from a device.

Reliability

Wireless sensor networks must be reliable. Reliability is achieved through:

- Secure communications
- Adapting to different types of interferers

- Retries on different paths, on different frequencies, and at different points in time
- Self organizing

The first point, secure communications, ensures that communications are not tampered with. The second point, adapting to different types of interferers, is what makes the network robust. In the wireless world interferers can be things like a truck parked in front of a transmitter, a cell phone, a walkie-talkie, a BlueTooth device, etc. To handle interferers WirelessHART is designed to retry on different frequencies, on different physical paths, and at different points in time. Since there is no way to predict all of the sources of interference using static site surveys, the sensor network must be self-organizing and must adapt. The only topology robust enough to support long term operations is a mesh. As more devices are added to the sensor network the mesh gets stronger and more robust. As we have seen, additional devices improve path diversity. Additional access points, and devices in general, increase throughput, reduce latency, and can be used to route around areas of potential interference.

9-4. Communication Rules

The network manager is responsible for managing routes, graphs, and allocating communication resources. Once these communication resources have been distributed to devices they are utilized by devices to send and receive communications. When contention for resources occur, priority is used to decide which messages get serviced first. The rules for communications are summarized in the following sections.

Network Routing

Three routing techniques are used to route packets: graph routing, superframe routing, and source routing. When using graph routing, a device sends packets with a Graph ID in the network layer header along a set of paths towards the destination. All devices on the way to the destination must be pre-configured with graph information that specifies the neighbors to which the packets may be forwarded. In a properly configured network, all devices will have at least two devices in the graph through which they may send packets.

Separate graphs are used to communicate from the device to the gateway and from the gateway to each device. In many cases a single graph can be used to communicate from devices up to the gateway. Individual downward graphs from the gateway to devices are only required in cases where devices such as control valves are being used to close loops.

Superframe routing is a special case of graph routing. In superframe routing packets are assigned to a superframe. Packets are instructed to follow the superframe en route from the source to the destination. With superframe routing the Graph ID is set to the Superframe ID. Since the packet follows the superframe, it is not necessary to explicitly configure graph edges.

With source routing, pre-configuration of the forwarding devices is not necessary. To send a packet to its destination, the source device includes in the network layer header an ordered list of devices through which the packet must travel. As the packet is routed, each routing device utilizes the next device address from the packet header to determine the next hop to use. Since packets may go to a destination without explicit setup of intermediate devices, source routing requires knowledge of the network topology. Only the network manager has enough information to construct source routes.

The network manager contains a complete list of routes, graphs, and devices. When devices are initially added to the network, the network manager stores all neighbor entries including signal strength information as reported from each device. The network manager uses this information to build a complete network graph. The network graph is optimized for reliability, hop count, reporting rates, power usage, and overall traffic flow (Figure 9-18).

Every graph in a network is associated with a unique Graph ID. To send a packet on a graph, the source device includes a Graph ID in the packet's network header. The packet travels along the paths corresponding to the Graph ID until it reaches its destination, or is discarded. In order to be able to route packets along a graph, a device needs to be configured with a graph table. The graph table contains entries that include the Graph ID and neighbor address. Redundant paths may be set up by having more than one neighbor associated with the same Graph ID. The device routing the packet must perform a lookup in the graph table by Graph ID, and send the packet to any of the neighbors included in the table. Once any neighbor acknowledges receipt of the packet (Media Access Control [MAC] level acknowledgement), the routing device may remove the packet from its transmit buffer. If an acknowledge is not received, the device will attempt to retransmit the packet at its next available opportunity. The packet will be retried until it is successfully sent or is timed out.

The network manager generates a set of graphs covering communications from each device to the gateway, from the gateway to each device, and broadcasting from the gateway to all devices in the network. Routes and graphs are allocated using a set of rules as follows:

- If possible one of the neighbors should be an access point (i.e., a single hop).
- Route through powered devices if they are available.
- Minimize the number of hops.
- Use a combination of weighted signal strengths to select between alternative routes (the weights need to be applied to all of the nodes in a particular set of hops towards the destination).
- Keep the number of hops to four or less.
- Prune the number of neighbors to four or less.

Network Scheduling

Network scheduling involves allocating superframes, links, routes, and graphs. During scheduling the network manager ensures that all the devices in the network are given sufficient network resources to meet their data reporting needs. As part of doing this the network manager also pays attention to latency, as well as power consumption. Managing latency, while at the same time conserving power, is an important consideration. To minimize latency, communications need to be organized so that packets are not delayed en route from source to destination. To minimize the energy use, devices should be placed into sleep mode as much as possible. Power consumption limits the utility of wireless sensor networks, which must operate unattended on the order of years. Replacing batteries is a laborious task and extremely difficult in some environments. Conserving energy is therefore critical for prolonging the lifetime of the sensor network.

WirelessHART takes the approach that applications and devices specify their communication requirements and a centralized network manager, in turn, allocates communication resources. The centralized network manager then distributes the schedule to each of the effected devices. When devices are not scheduled to transmit or receive they sleep.

Each wireless network contains exactly one overall schedule that is created and managed by the network manager. The schedule is subdivided into superframes. Each superframe is further subdivided into superframe relative links that repeat as the superframe cycles.

Requests for communication resources are directed to the network manager through network management commands. For example, a software application may request a device to schedule a burst mode communication every second. The device in turn would make a request to the network manager for the communication resources that it needs to meet this request. This kind of network-centric approach allows a sensor network to collect and process data in a very energy-efficient manner.

Over the lifetime of the wireless network it is important that the network manager be able to automatically adapt the schedule. For example, applications like asset management are started and stopped, valve and device diagnostics are run, devices are added and removed, and interferers such as scaffolding or other communication devices come and go. As part of the adapt algorithm, the network manager continually receives feedback. Feedback is provided through periodic health reports, network level alerts, and network statistics. Adapting includes adding and removing superframes, links, graphs, and routes. Summarizing the key points that have been presented:

- Scheduling reduces contention and ensures that all devices are provided with the opportunity to communicate. This is one reason why network scheduling yields such high end-to-end packet reception.
- Scheduling must be power aware.
- In many situations broadcasting is used to improve the overall effectiveness of the network. The network manager must set up routes for broadcasting.
- Time synchronization is critical—the entire network is dependent on a common understanding of time.
- To minimize latency the network manager needs to allocate links in sequence from the source to the destination.

- The sensor network must adapt to changing conditions in the plant and in the software applications.

Priority

Being able to handle sudden increases in traffic is an important characteristic of all networks. In wireless networks one of the key mechanisms used to do this is message priority. Message priority helps mitigate situations where devices need to perform information flow control. An example of this is when a piece of process equipment trips, for example a boiler. In these situations it is important that measurement and control communications get through while some of the less time-critical information is delayed. Priorities are set so that network management commands are the highest priority (this allows the network manager to temporarily increase communication resources, re-route, etc.), followed by real-time published messages, followed by alarms and alerts. In the case of alarms and alerts, since they are time-stamped at the device, no critical information is lost if an alarm or alert is delayed a few time slots.

Optimizing Communication Resources

Scheduling is performed by a centralized network manager which uses overall network routing information in combination with communication requirements that devices and applications have provided. The schedule is subdivided into links and transferred from the network manager to individual devices; devices are only provided with the links for which they have transmit or receive requirements. The network manager continuously adapts the overall network to changes in network topology and communication demand.

In order to meet the requirements for latency and jitter, process measurements must be communicated often enough to not miss oscillations in the signal. To support doing this in a very efficient manner, measurements can be published periodically or periodically-by-exception—when reporting by exception a maximum time interval can be specified [2]. In addition, the reporting rates can be adjusted to report more frequently when the signal value is higher or lower than a configured threshold value. Measurements also include their sampled time stamp. Multiple variables can be communicated at the same time.

The wireless update time termed "default update rate" (refresh time) is the time interval for periodic reporting. The wireless update sensitivity termed "trigger level" is the minimum change in measurement value for exception reporting. To save power, the transmitter goes to sleep and wakes up periodically to check if the change in sensor value from the last value transmitted is larger than the trigger level. When the change exceeds the trigger level or the time since the last communication exceeds the default update rate, then the value is transmitted. The time interval between periodic checks of the sensor is the "triggered update rate" (wakeup time). If only periodic reporting is available, this time interval is the default update rate. Increases in the default update rate and trigger level settings reduce the number of transmissions, which increases battery life. The PIDPlus eliminates the ramps, limit cycles, and spikes from large values of these settings facilitating an increase in battery life.

Monitor and Control over a Wireless Sensor Network

Wired digital communications have been in use for many years in supporting monitor, control, and diagnostic applications. Most plants today use digital communications for a significant portion of their control applications. Wireless digital communications are relatively new in process plants. One of the key questions that is often asked is, "Is wireless ready for control applications [2]?"

WirelessHART is effectively HART and as such was designed from the beginning to support monitor and control applications. Key factors such as sampling intervals, jitter, and latency are either easily managed or are well within the requirements of typical process control applications. To minimize communications, exception reporting was added. The rule of thumb is that feedback control should be executed four to ten times faster than the process response time, where response time equals the process time constant plus deadtime. Because wired measurement systems are often unsynchronized with the control system (Figure 9-7), measurement values are oversampled by a factor of as much as two to ten times.

With wireless systems, because of the inherent time-synchronized nature of the mesh network, it is possible to reduce or eliminate this oversampling. Sampling can be scheduled to occur just before communications are scheduled. It is also possible to turn on exception reporting and only send values that have changed (Figure 9-8).

A control system's ability to meet its control performance requirements is affected by both delays (latency) and variation (jitter) in when the information is available. Latency and jitter are often introduced by the commu-

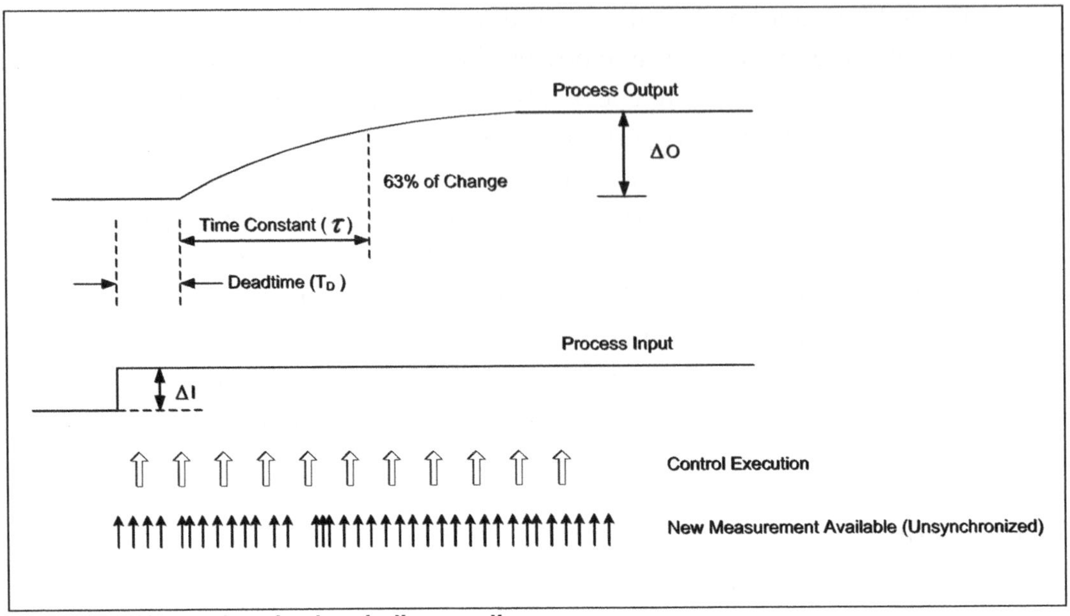

Figure 9-7. Unsynchronized periodic sampling

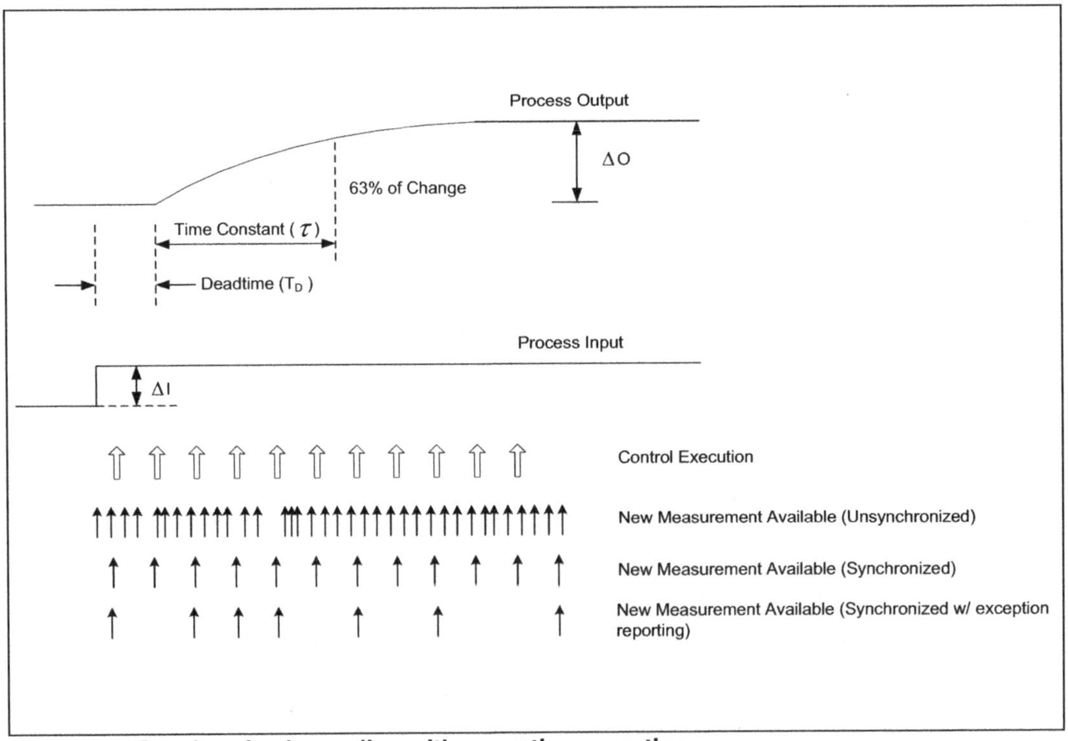

Figure 9-8. Synchronized sampling with exception reporting

nication system—for example, when data is transmitted from a transmitter to a gateway. Latency is the time it takes for a communication packet to make its way from the source to the destination, while jitter is variation in latency between different communication cycles (Figure 9-9). Excessive latency (which effectively adds deadtime to the process) and jitter (which adds error into the control calculations) can lead to significant degradation in control performance.

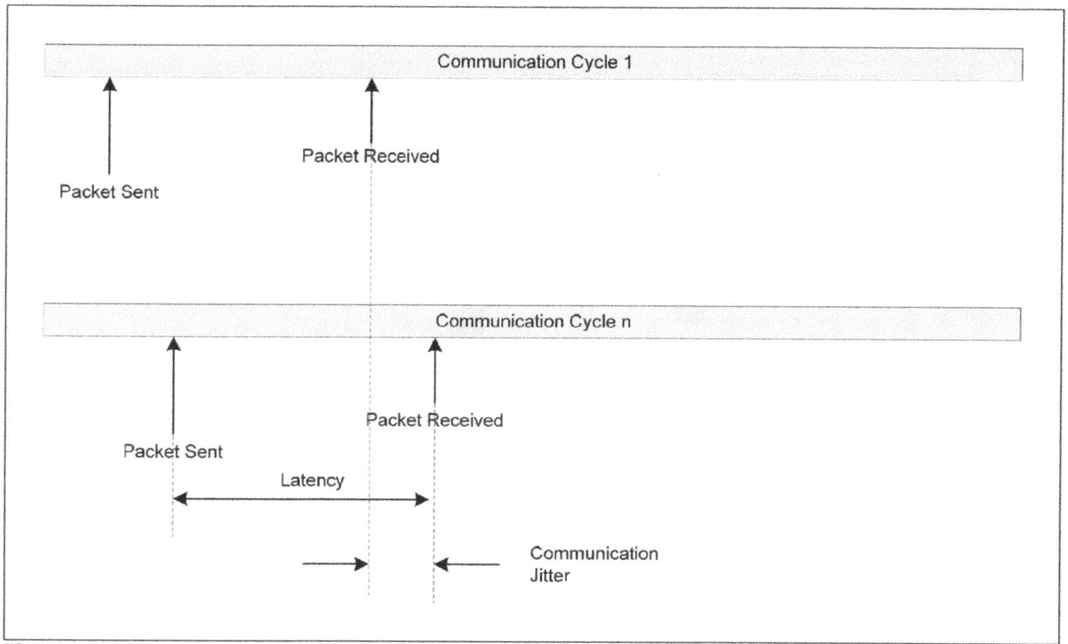

Figure 9-9. Latency and jitter introduced by communication cycles

Comparing wireless to wired, wireless transmission rates are faster than some traditional wired fieldbus technology. For example, if the communication rate is 31.25 kilobits/second, the communications delay will be 32 microseconds/bit. Wireless has a much faster communications rate—250 kilobits/second—so the delay introduced by the communications rate is only 4 microseconds/bit. Since communications occur in 10-ms time slots (Figure 9-10), the time delay for a single hop will be 10 ms.

In many scenarios communications require more than one time link for a message to travel from the source to the destination (Figure 9-11). If a communication can't reach its destination directly, it can "hop" from device to device. This ability to route around physical obstacles or interference is a core feature of mesh technology.

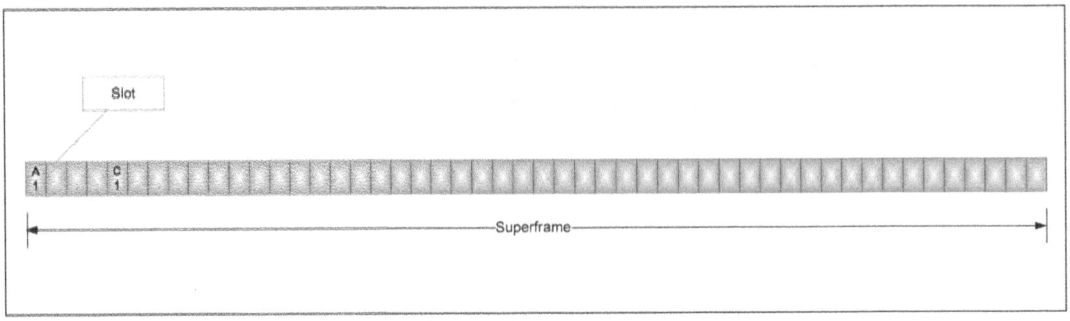

Figure 9-10. Time links for source to destination

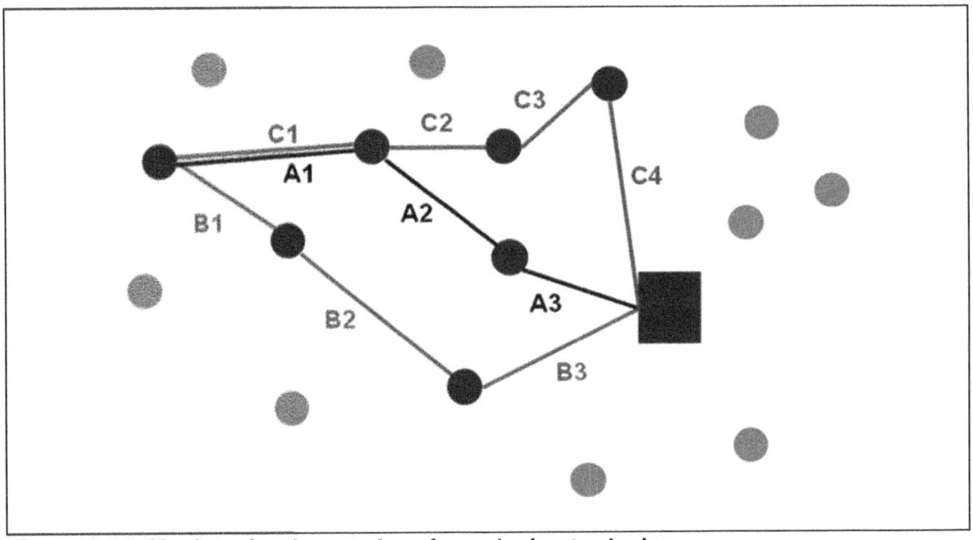

Figure 9-11. Mesh technology to hop from device to device

Changing the route the data travels can contribute to variations in communication time (jitter). Although each additional hop increases latency, in typical applications the average delay is well within the requirements for control.

In most cases, a wireless network will be able to retry a failed message in the next time slot or the one following. For example, we'll assume it takes 10 ms to process a message and assign it to another time slot. Path A in Figure 9-11 could therefore produce as much as 50 ms of total latency (10 ms + [10 ms + 10 ms] + [10 ms +10 ms]). Path B has the same number of hops and thus the same communications latency. But Path C has an additional hop, bringing total communications latency to 70 ms. This timing difference introduces a 20-ms jitter in the communications. (In many cases, the routing device will be able to retry in the very next link, which would reduce the total latencies to 30 ms for Paths A and B and to 40 ms for Path C.)

Experience in hundreds of wireless field device installations shows that communication latency on average is much lower than in this example. In real plant settings, typically 30% of the devices communicate directly with the gateway or network access point (10 ms) and about 50% are one hop away (30 ms). The rest are three to four hops. Using these numbers from existing plant installations, the average latency time will be about 30 ms.

For actuators, interlocks, or any device that affects the process, the network manager is responsible for the establishments of routes, graphs, superframes, and links that the control system may use to send set points and outputs to the user layer application in field devices.

Network Management and Host Request

The first step in the formation of the wireless network is to allocate communication resources for network management. Devices wishing to join the wireless network listen for advertisement packets and then use information in the packets to generate join requests. When a join request is received by the network manager, the network manager verifies the device's join key [5] and either completes or rejects the join process. As part of this process, the network manager creates dedicated links in one of the superframes for device management and advertisement functions. A link will also be defined for the newly added device to send advertisement packets. This new link information is transferred into the network and reflected in all of the effected devices.

It is expected that any new device will listen for a period of time for advertisement packets and select the device with the strongest signal for joining the network. If two new devices happen to transmit a join request at exactly the same time using the advertised link and channel, then the messages will collide. In this event, the collision will be detected and the transmission re-attempted after a backoff time. Thus, the join requests will be spaced and will most likely be successful on the next attempt.

The overall network routing will be determined by the network manager based on signal strength as determined by the physical layout of the plant, number of hops, and traffic flow. The overall routing will be further adapted to reflect diagnostic information, retries, etc.

To support the configuration of multiple superframes for the transfer of process information at different rates, the measurement scan rates used by the field device and the associated communication should be configured as integer multiples of the fastest update time that will be supported by field devices (an example of a field device is a pressure transmitter). For this example, the supported scan rate will be defined as 2^x where x is positive or negative integer values; for example, scan rate selections of 1 sec, 2 sec, 4 sec, 8 sec, 16 sec, and 32 sec.

To avoid introducing latency into the measurement value that is communicated to the gateway, it is important that processing by the sensor be coordinated with the link configured for the measurement transmission (Figure 9-12).

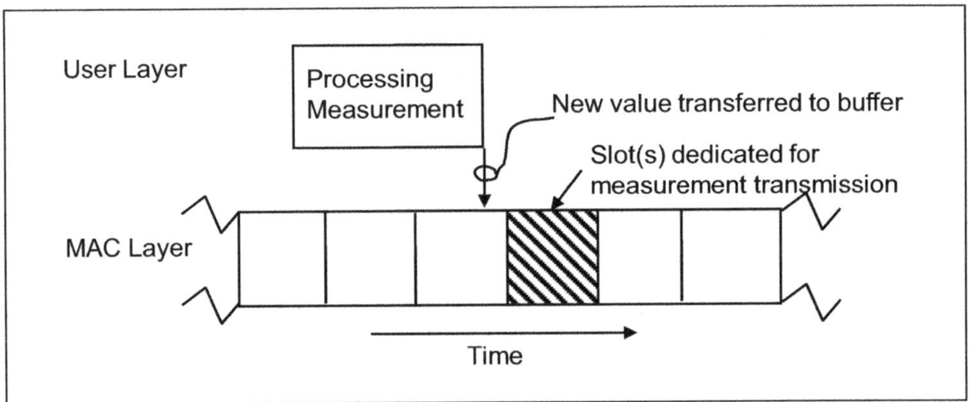

Figure 9-12. Synchronization of measurement processing and transmission

The scheduling of communications associated with process measurements included in a network can be simplified by defining a superframe for each scan period and developing the schedule by allocating links for transmission of measurement data starting with the fastest to the slowest scan rates. In the schedule, a device may only occur once within a link time since at any given time a device is either transmitting or receiving on one channel (i.e., since each device has only one radio, it can only use one channel at the same time).

Scheduling Example – Single Hop

As an initial example of scheduling the most simple case will be used: all devices are able to communicate to the gateway through a single hop.

Using the recommended approach of defining superframes for each scan period and allocation of links from the fastest to the slowest measurement, the graph associated with the batch column, reactor, and vessel measurements would appear as a hub-and-spokes diagram (Figure 9-13). Routing information would be shown as a set of single hops (Figure 9-14).

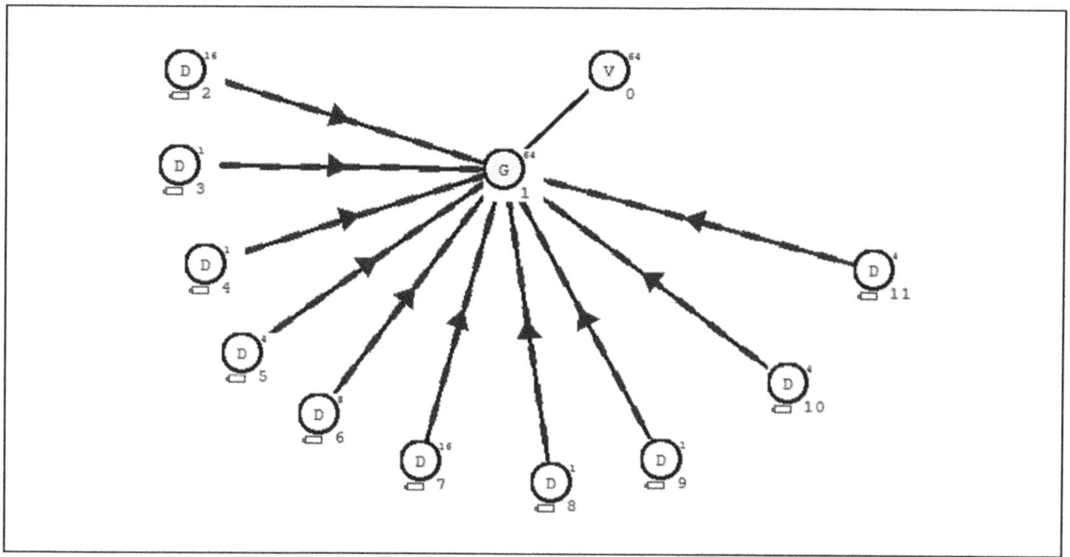

Figure 9-13. Batch column, reactor, or vessel example – single hop

To support immediate re-transmission after a failed transmission to the gateway, additional links would be added in the schedule immediately after each transmission. The schedule for this configuration is illustrated in Figure 9-15. The schedule shows radio utilization in the top portion and device communications in the bottom portion.

Scheduling Example – Multiple Hop

The previous example presented the ideal case where all network devices were directly connected to the gateway. In most networking scenarios multiple hops will be used (Figure 9-16).

In multi-hop mesh networks, the routing is through two or more neighbors. At the point of communication, the device decides which neighbor to send to based on the link that is available. For example, device 2 can send packets through either device 4 or device 5 (Figure 9-17).

The schedule for this multi-hop example is shown in Figure 9-18.

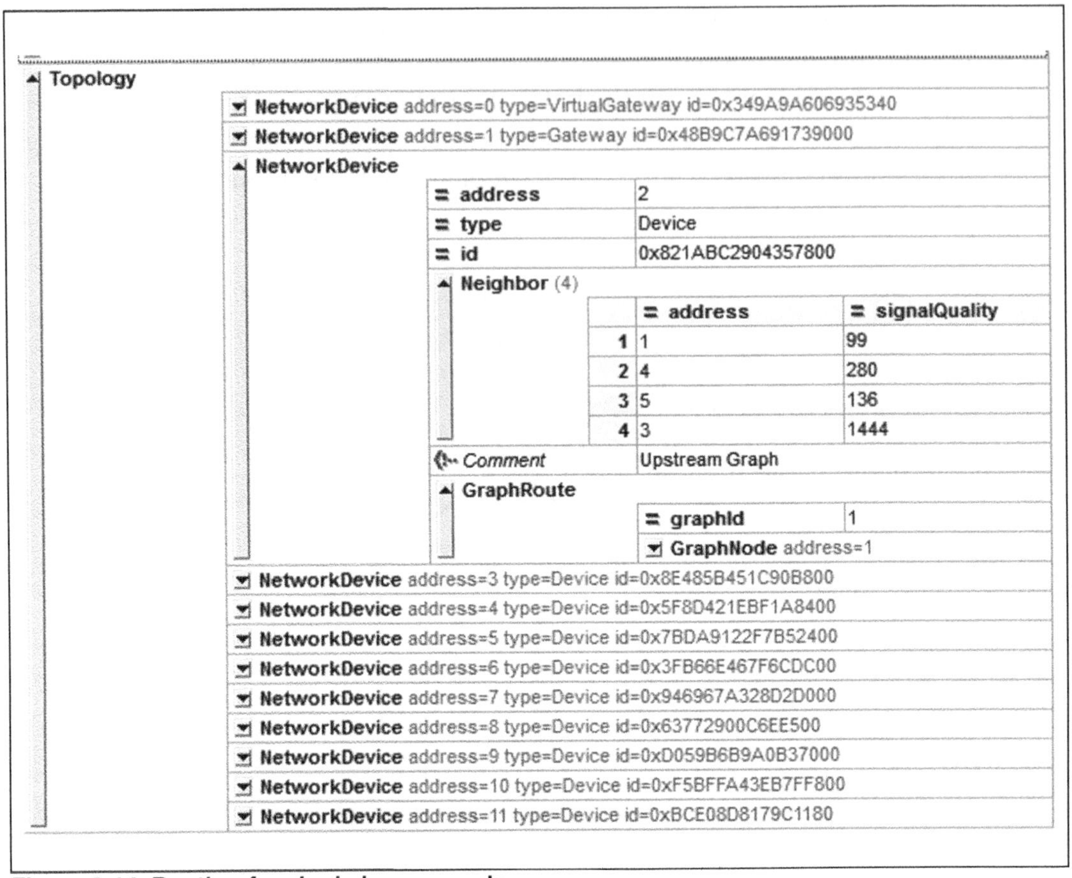

Figure 9-14. Routing for single hop example

Scheduling Example – Incorporating Control Devices

The example up to this point has only shown measurement devices. As a final step in this example, all the throttling valves are included at 1s scan rates, and all the blocking valves are included with 4s scan rates (Figure 9-19).

The example also shows the effects of adding a second access point (shown as G1 and G14 in the diagrams). The overall schedule is shown in Figure 9-20.

Over time the network topology will change for example a truck could drive into the middle of the mesh. When this happens the network manager will restructure the network—this is illustrated in Figure 9-21.

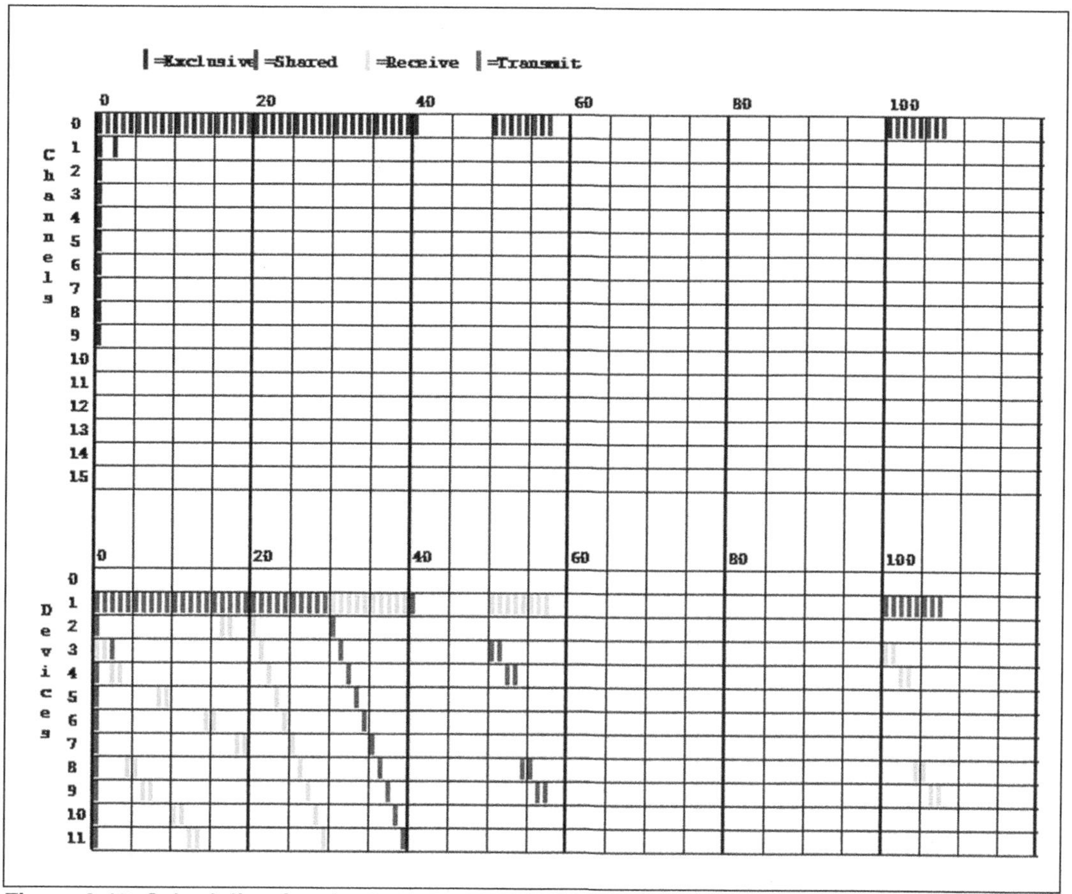

Figure 9-15. Scheduling for single hop example

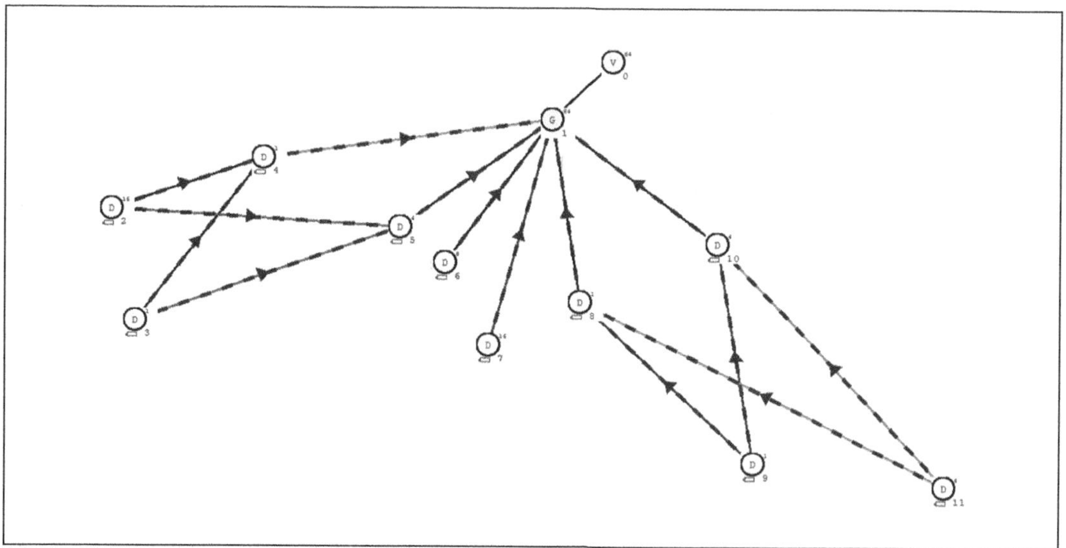

Figure 9-16. Batch column, reactor, or vessel example multiple hop

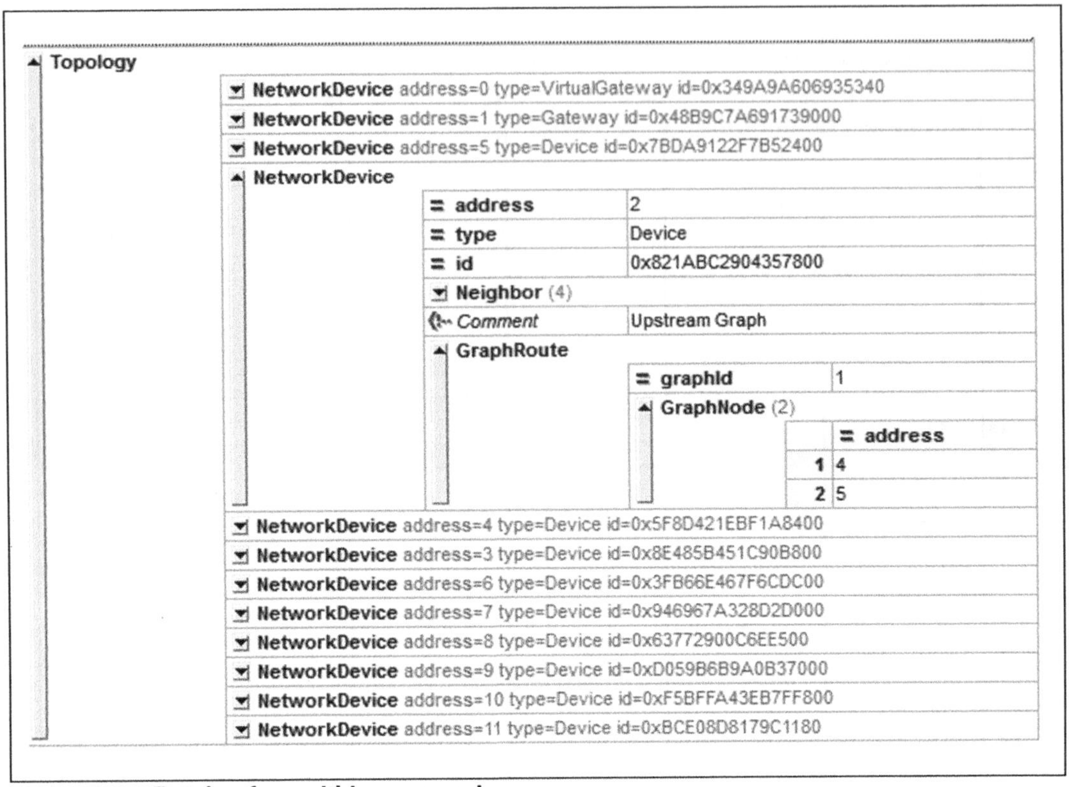

Figure 9-17. Routing for multi-hop example

9-5. Process Control

Utilizing wireless communication to provide measurements that will be used in closed loop control was illustrated earlier in Figure 9-7 and Figure 9-8. As discussed, power consumption can be reduced by eliminating oversampling and by only communicating values when they change.

A modern wired process control system is often structured as shown in Figure 9-22 [21][24]. As illustrated in this figure, sensors and actuators are connected to controllers via control networks, such as Fieldbus [15] and Profibus [17]. A sensor provides measurements and status of a physical property; for example, the flow in a pipe associated with the process. Based on the measurement from sensors, the controller determines any adjustment in the actuators that is needed to maintain the process at a target value, i.e., the set point. The control loop is executed periodically at a rate fast enough to correct any unwanted deviations in the process.

In contrast to non-wireless control systems, communications in a wireless control system tend to have more variation in communications. This variation may be caused by interferers, power failures, and environmental

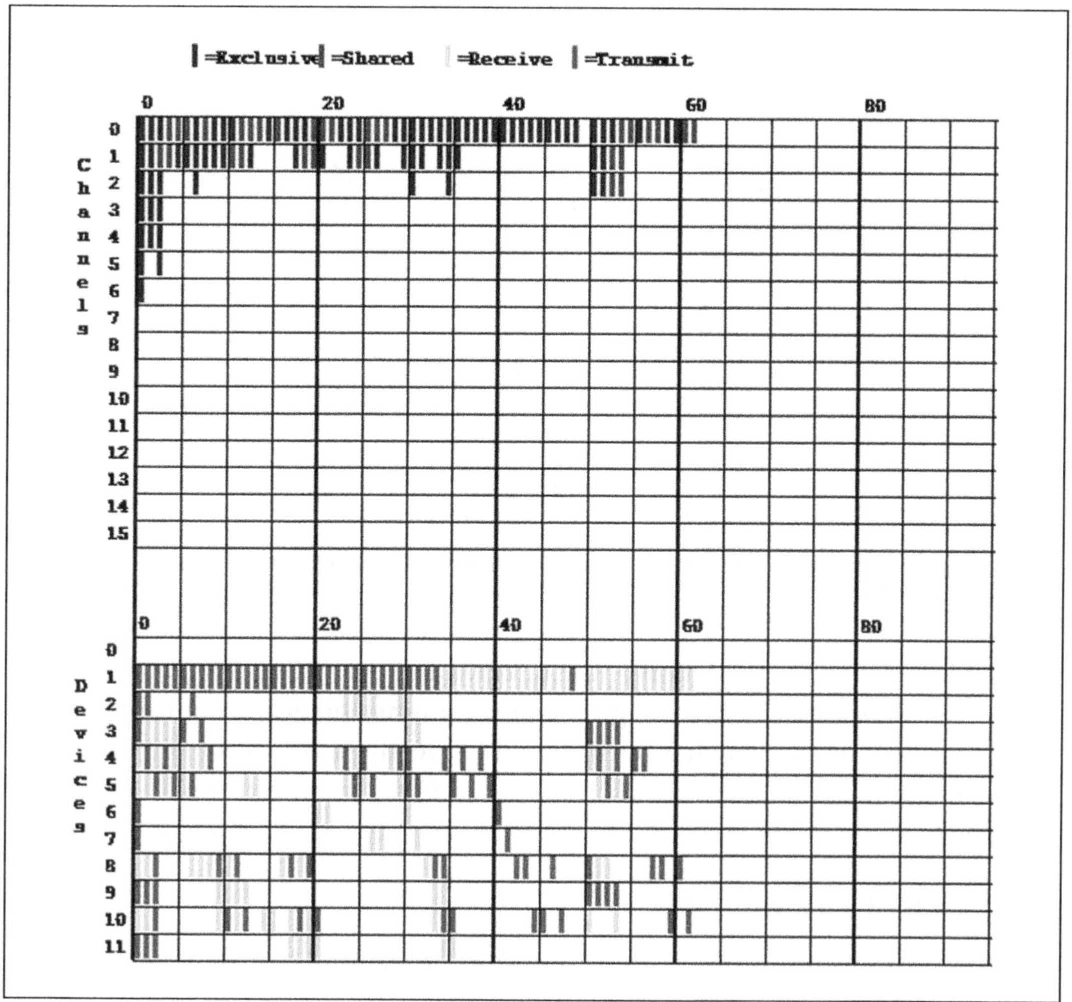

Figure 9-18. Scheduling for multi-hop example

factors such as lightning storms. Current control strategies address this by oversampling. In standard process control systems, a missed I/O communication is considered an error. Usually, a control loop is configured with a maximum number of lost I/O communications, after which the loop declares failure and the values are set according to the fail-safe configuration.

Therefore, communication variation presents a challenging problem for standard control paradigms. Consider a PID block with inputs from a wireless channel. Suppose the inputs are lost at time t_1 and re-established at time t_2. The derivative component of the PID would cause a spike in the output at t_2. Also, from t_1 to t_2, the reset component may wind up based on the error that existed at time t_1.

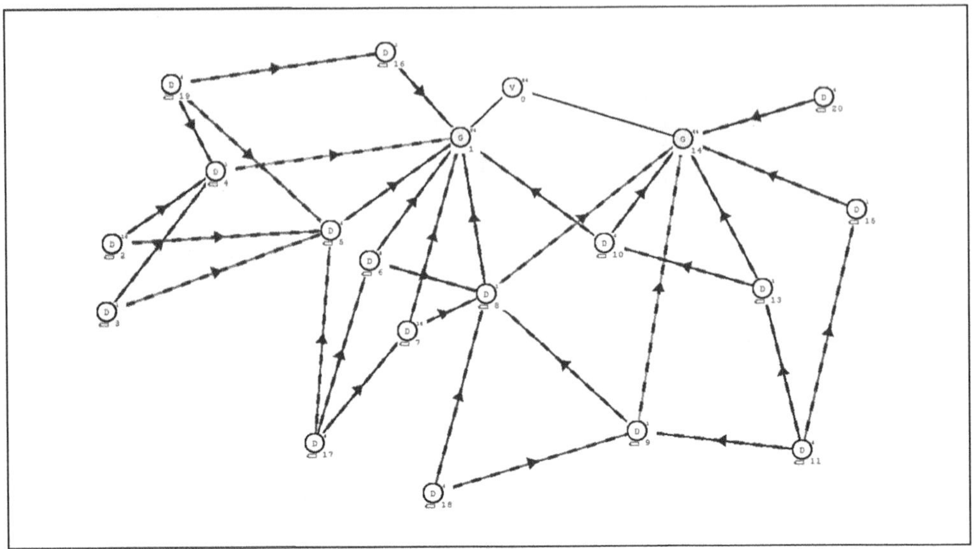

Figure 9-19. Batch column, reactor, or vessel example – all valves and devices included

Modifying the calculation of the integral and derivative components of PID algorithms by detecting missed communications and compensating for them proactively has been shown to greatly improve control performance [2, 3]. This section includes a significant portion of a paper published by ISA [3].

The Standard PID Algorithm

PID is the most widely used control algorithm in industrial process control [25]. As shown in Figure 9-23, the controller compares the process variable (PV) with a reference set point (SP). The error is then processed to calculate a new output to bring the PV back to its desired SP [26].

PID stands for "proportional, integral and derivative" components of the algorithm. Each of the three components performs a different task and has a different effect on the functioning of a system. Their outputs are summed to produce the system output.

Though there are many variations of PID algorithms, in its non-interacting form without rate limiting and all actions based on error, the equation for the standard PID algorithm is

$$Output = K_P\left[e(t) + K_I\int e(t)d(t) + K_D\frac{de(t)}{dt}\right]$$

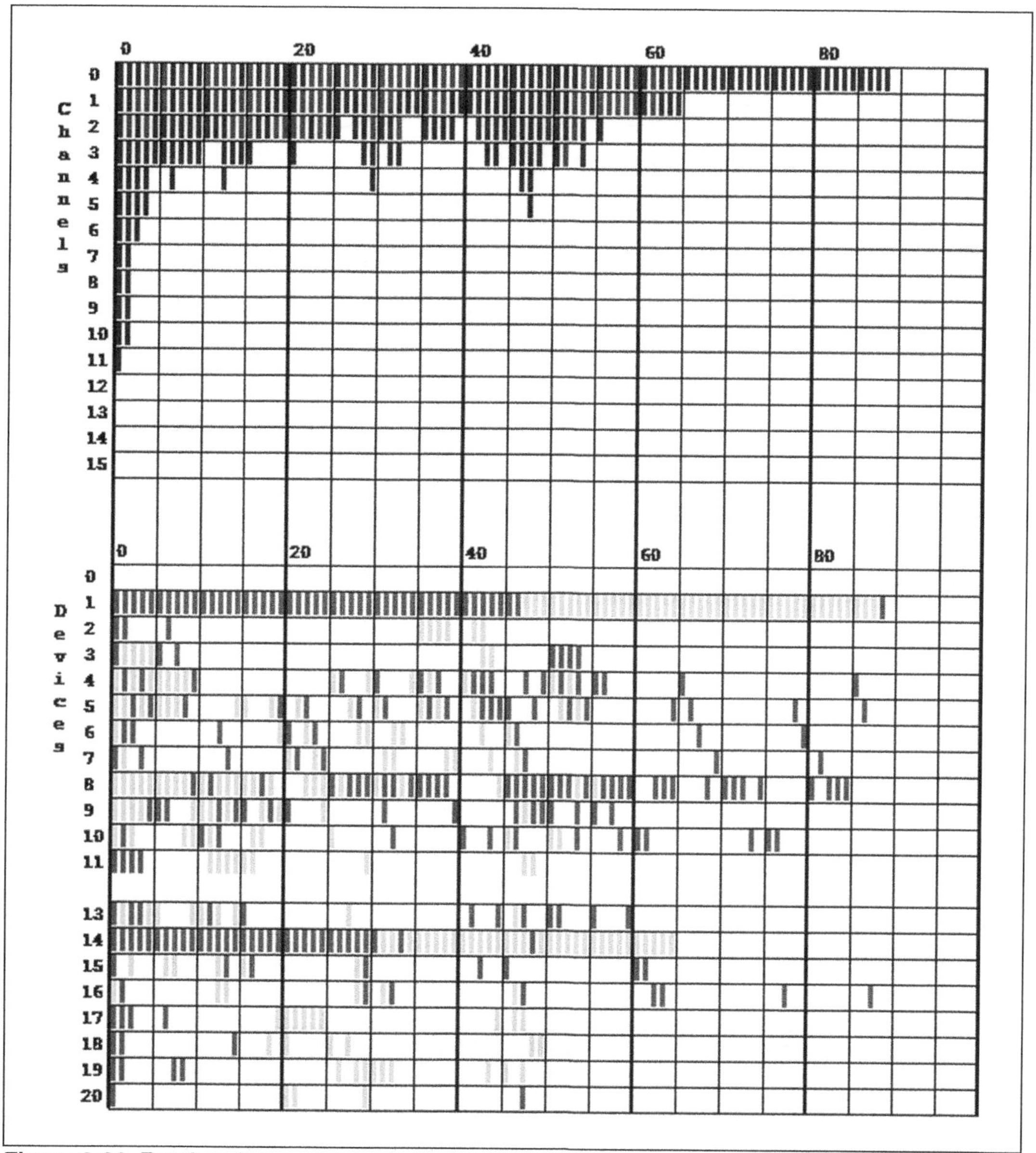

Figure 9-20. Batch column, reactor, or vessel example – schedule

K_P, K_I and K_D are the proportional, integral, and derivative gains, respectively.

In its digital form, the software implementation of the PID algorithm is based on the sampled data for its PV being provided on a periodic basis.

When there is no communication loss, PID, once configured for the process it controls, keeps the process in a steady state. Figure 9-24(a) shows the PID reaction to a process disturbance.

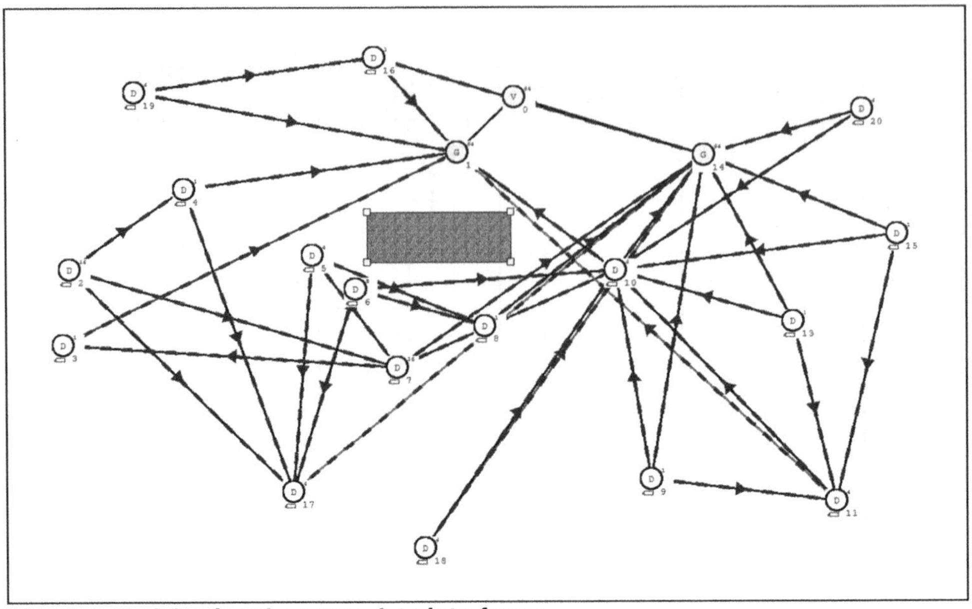

Figure 9-21. Adapting the network to interferers

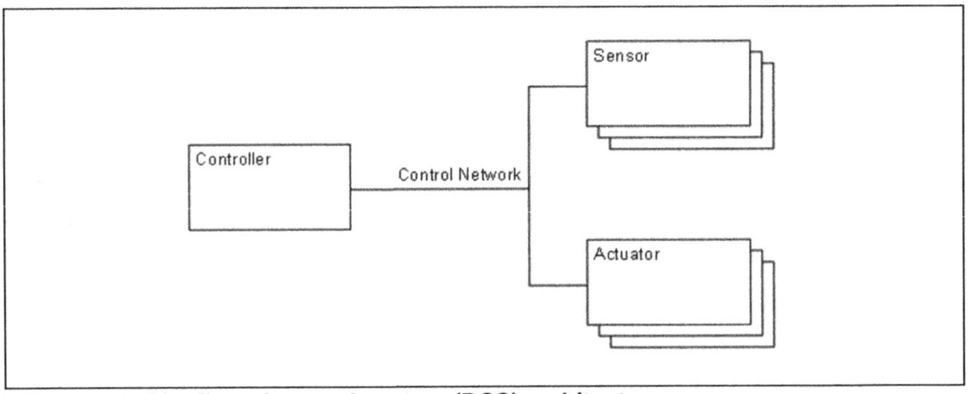

Figure 9-22. Distributed control system (DCS) architecture

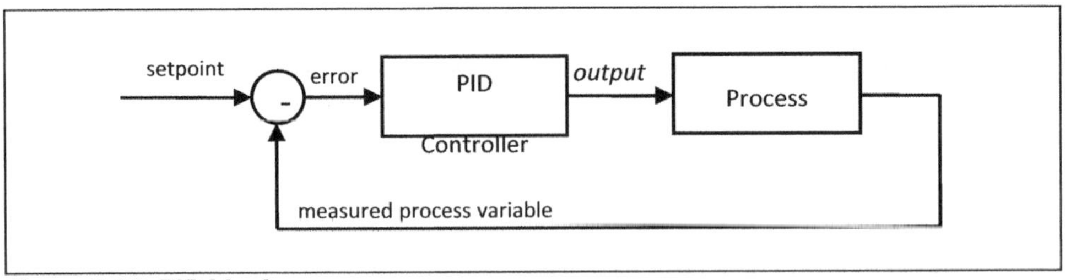

Figure 9-23. PID block

Before time t_0 the PID output ("out") is kept at a constant value to maintain PV at the SP value. At t_0 a drop of PV is observed due to process disturbances. To correct the drop, the PID increases its output. At time t_2, PV comes back at SP, and the out is stabilized at some value that is slightly bigger than its original value to compensate for the disturbance. As is shown in Figure 9-24(a), the out is the sum of three parts: P, I, and D.

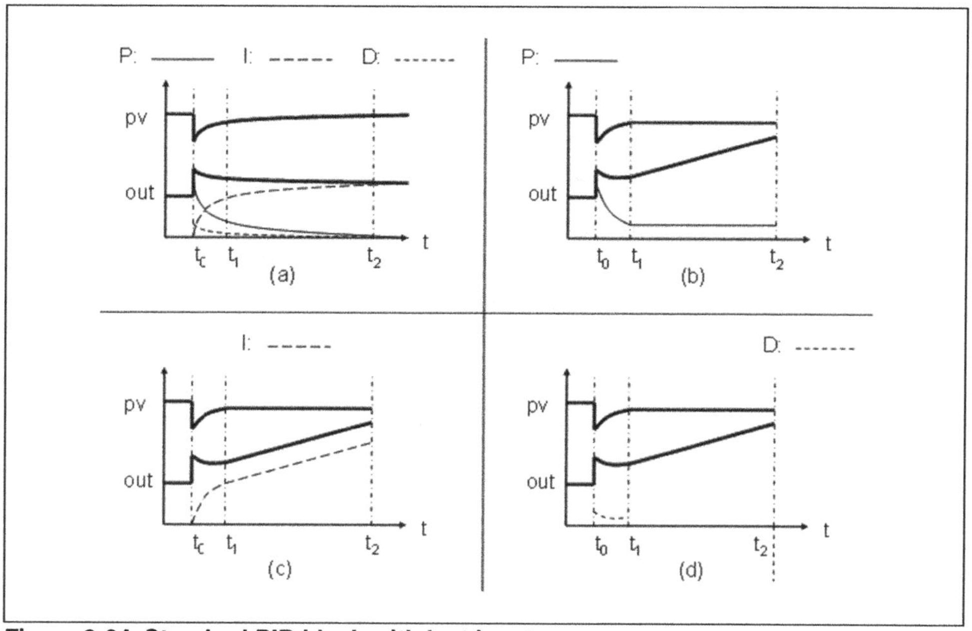

Figure 9-24. Standard PID block with lost input

Input Communication Lost

Now consider what would happen to each component of PID output if the communication from the sensor input is lost between times t_1 and t_2. For the PID block, the measured PV remains the same as at time t_1 during this period, as shown in Figure 9-24b, c, and d.

Figure 9-24(b) shows the proportional gain P. Since the measured PV and SP remain the same, the proportional gain is constant from t_1 to t_2. Figure 9-24(c) shows the integral part I. Since PV and SP remain the same, the error remains the same, so the integral part is a linear increasing line from t_1 to t_2. Figure 9-24(d) shows the derivative part D. Since PV and SP remain the same, the error remains the same, so the derivative stays 0 from t_1 to t_2. As a result, the out of the PID block is a linear increasing line from t_1 to t_2, shown in Figure 9-24b, c, and d. This destabilizes the process. The longer the communication is lost, the bigger the deviation between PV and SP.

Once the communication is reestablished at t_2, PID is back to normal. However, since the derivative part calculated at time t_2 is based on the difference of measured PVs between t_2 and one period before t_2, we shall expect a spike for the derivative part. This is because the PV at t_1 is used as the PV at one period before t_2. PV at t_2 could be significantly different from the PV at t_1. Since the PV changes between time t_1 and t_2, we expect the derivative spike to be even bigger. Due to sudden changes of the proportional and derivative parts, the value of the out will have a big impulse before and after t_2.

Output Communication Lost

We continue to analyze how standard PID behaves if output communication is lost between t_1 and t_2. Here we assume there are no other disturbances and input communication is OK.

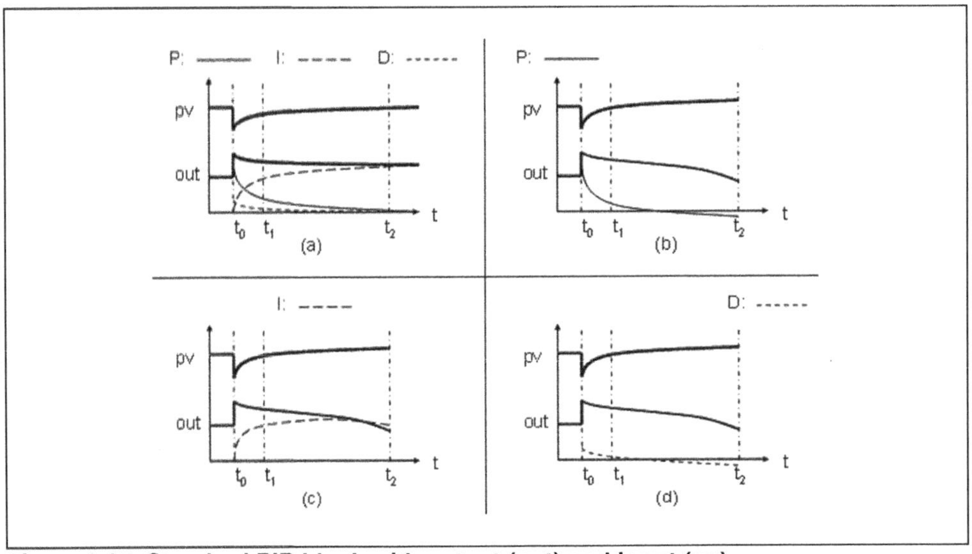

Figure 9-25. Standard PID block with output (out) and input (pv)

From t_1 the actuator will stay with the out value of t_1 until t_2 when a new out value is received from the PID. This will cause the measured PV to eventually reach SP and then overshoot a little bit, shown in Figure 9-25 (b,c,d). The P, I, and D components are all calculated based on the current PV, shown in Figure 9-25 (b,c,d). This is as good as we could expect from PID. The only drawback is that the actuator gets a bump in the out value, from the one calculated at t_1 to the one calculated at t_2.

Both Input and Output Communication Lost

When both input and output communications are lost, the PID behaves the same as when only input communication is lost, as in Figure 9-24. The only difference is that when communication is reestablished at t_2, the actual PV is different. In this case, the PV strays less as the actuator output stays constant. Similarly, the actuator receives a bump in the out value.

The Enhanced PID Algorithm

The underlying assumption in the digital implementation of the PID algorithm above is that the algorithm is executed on a periodic basis. When the input containing the measurement is lost, the calculated reset action may not be appropriate [14]. When later on a new measurement gets through, the calculated derivative action may produce a spike in the output. If a PID block continues to execute using the last process variable, the output will continue to move based on the reset tuning and error between the last measured process variable and the set point. If the control block is only executed when a new measurement is communicated, then this could delay control response to set point changes and feedforward action on measured disturbances. Also, when control is executed, calculating the reset contribution based on the scheduled period of execution or on the time since the last reset contribution may result in changes that increase process variability [14].

In [14], an enhanced PI algorithm was proposed to reduce wireless communications between sensors and controllers without significantly impacting control performance. Based on that work, further improvements to the derivative part of the equations were made and applied [3].

To provide the best control when measurements are not updated on a periodic basis, the PID may be restructured to reflect the reset and derivative contributions for the expected process response since the last measurement update. One means of doing this is illustrated in Figure 9-26.

As shown in Figure 9-26, the reset/rate contributions (integral/derivative parts of the PID) are determined based on the use of a new value flag from the communications stack, the same idea as in [14]. To account for the process response, the filter output is calculated in the following manner when a new measurement is received:

$$F_N = F_{N-1} + (O_{N-1} - F_{N-1}) * \left(1 - e^{\frac{-\Delta T}{T_{Reset}}}\right)$$

$$\begin{aligned} where \quad F_N &= New\ filter\ output \\ F_{N-1} &= Filter\ output\ for\ last\ execution \\ O_{N-1} &= Controller\ output\ for\ last\ execution \\ \Delta T &= Elapsed\ time\ since\ a\ new\ value\ was\ communicated \end{aligned}$$

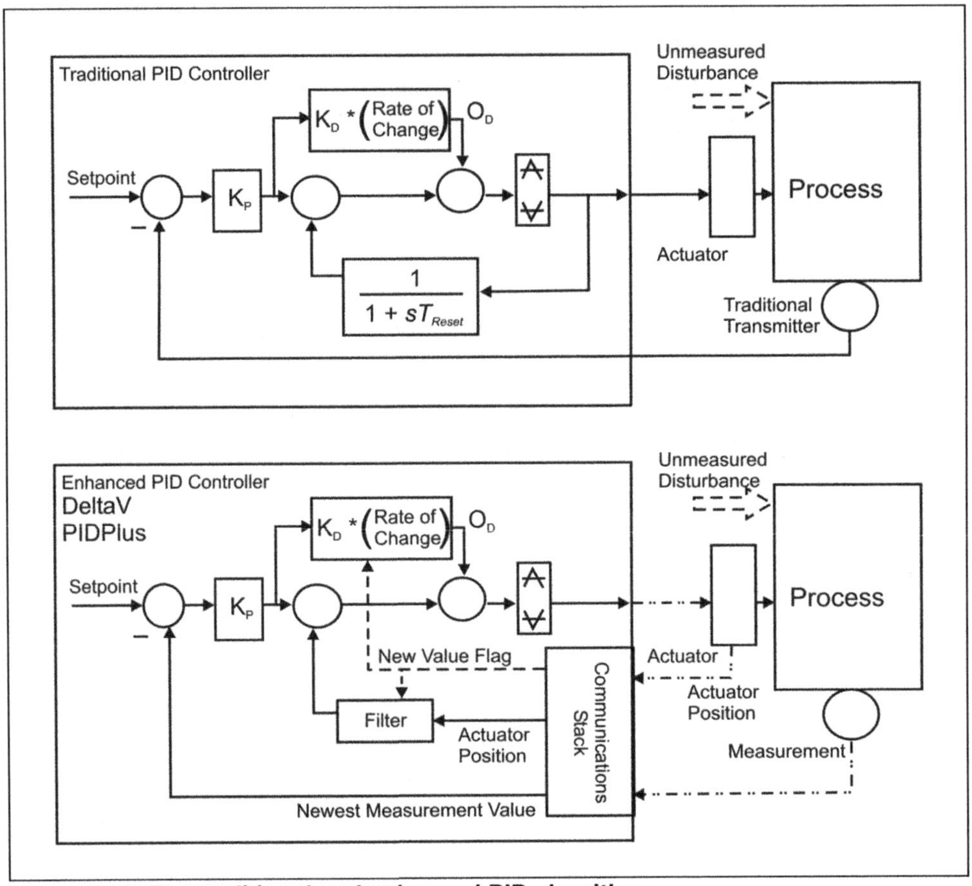

Figure 9-26. The traditional and enhanced PID algorithms

Since the last communicated actuator position as reflected in the feedback of actuator position is used in the integral calculation, this automatically compensates for any loss in the output communicated to the downstream element.

The derivative part for this example (rate limiting not applied) is determined by the following equation:

$$O_D = K_P \bullet K_D \bullet \frac{e_N - e_{N-1}}{\Delta T}$$

where:

e_N = current error

e_{N-1} = last error

ΔT = elapsed time since a new value was communicated

O_D = controller derivative term

Consider the contribution of the derivative part when the inputs are lost for several periods. When the communication is reestablished, $e_N - e_{N-1}$ in the equation above would be the same for the original and modified algorithms. However, for the standard PID algorithm, the divisor in the derivative part would be the period, while that in the new algorithm is the elapsed time between two successfully received measurements. It is obvious that the modified algorithm would produce smaller derivative action than the standard PID algorithm.

There are two major problems with the standard PID algorithm when dealing with communication losses: continued execution during communication loss and sudden output change when communication is reestablished. The enhanced PID algorithm solves these problems by only computing the integral and derivative components when communication is established and incorporating actuator feedback into the reset calculation.

To summarize, the enhanced PID algorithm has the following beneficial features:

- PID integral mode is restructured to provide integral action to match the process response in the elapsed time (reset time is set equal to process time constant)
- PID derivative mode is modified to compute a rate of change over the elapsed time from the last new measurement value
- PID reset and rate action are only computed when there is a new value

The enhanced PID integral calculation is done only when there is an update to eliminate the continual ramp of the traditional PID, which uses old and possibly erroneous information about the PV.

The enhanced PID derivative calculation uses the elapsed time between updates to eliminate the spike of the traditional PID, which uses module execution time as the divisor.

Experiments and Results

Several experiments were carried out to validate this algorithm. The setup for the test is shown in Figure 9-27. PROC_1 and PROC_2 are two identical processes, each of which consists of a second order process with a delay of one second and time constants of six seconds and three seconds.

The modified PID algorithm is implemented in PIDPLUS, and PID2 is a standard PID block. The parameters for PID2 are determined by testing it with a tuning application, which suggests a gain of 0.85, a reset of 10.71, and a rate of 1.71. Then the tuning parameters of PIDPLUS are set the same as PID2. PIDPLUS is configured to utilize the BKCAL_IN value for the reset components.

The process variable communication is simulated by the COM_IN_1 and COM_IN_2 blocks, which are controlled by COM_STATUS_IN. If COM_STATUS_IN is set to 1, blocks COM_IN_1 and COM_IN_2 relay measurements accurately. Otherwise, the two blocks drop the measurements. The same logic is applied to the output using COM_OUT_1, COM_OUT_2 and COM_STATUS_OUT.

By changing the external set point and introducing some disturbances that impact each PID and associated process equally, the performance of the two PID blocks can be evaluated. The performance of the two PID blocks is collected in the PERFORMANCE block. The metric used is Integral Absolute Error (IAE). The scan rate for all blocks is set to 0.2 second. Initially, the uncontrolled disturbance (DISTURBANCE) to the processes is 20.

Reliable Communications

To simulate the case of reliable communications, both COM_STATUS_IN and COM_STATUS_OUT were set to 1. SP was changed from 50 to 60. The result is shown in the left part of Figure 9-28 (from time 11:10 to 11:11).

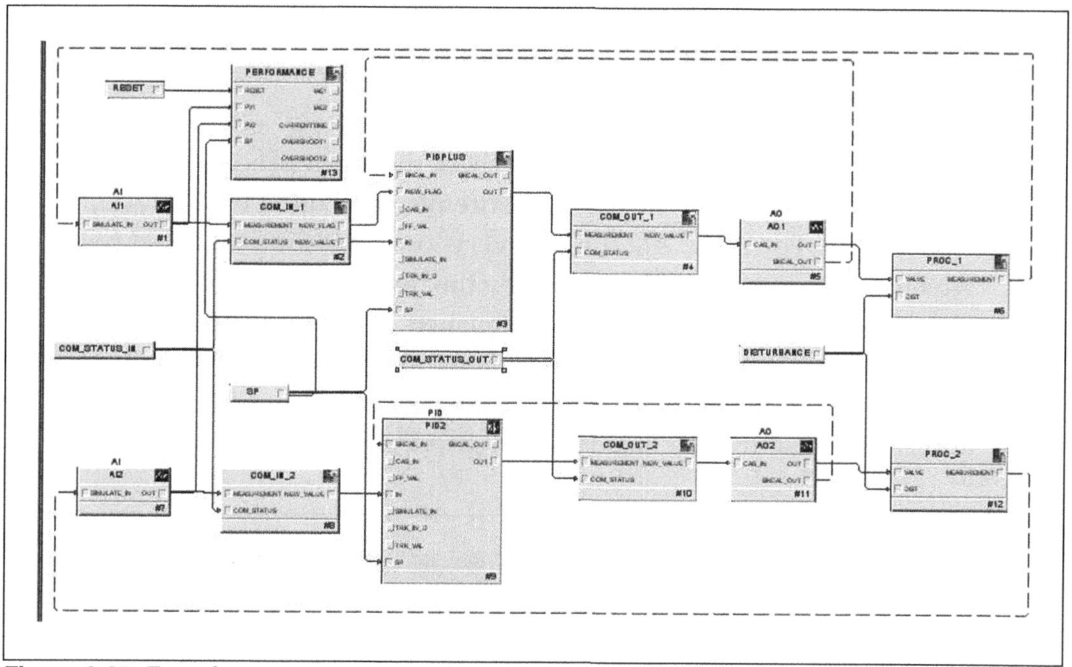

Figure 9-27. Experimental setup

The curve of AI1/OUT.CV matches nicely with that of AI2/OUT.CV, and AO1/SP.CV matches nicely with AO2/SP.CV. Therefore, we conclude that the two PID blocks work in the same way when communication is reliable.

Unreliable Communications

To study the effect of unreliable communications on the two PID blocks, both unreliable input and unreliable output are considered.

Unreliable Input

During the period of lost inputs, the last communicated process variable is maintained and used in the PID blocks. As a first experiment set points were used to introduce changes. The result is shown in the right part of Figure 9-28, where SP was decreased from 60 to 50.

When the input channel is shut down, the error between the set point and process variable fed to PID blocks is a constant. For the standard PID block PID2, the integral part would keep integrating, which explains the linear decreasing of AI2/OUT and AO2/SP. However, as it has a flag for missed communications, PIDPLUS would simply freeze the reset component during loss of communication, which explains the level-off in AO1/

SP during lost communications. Since AO1 is given a constant value, the process variable of PROC_1 approaches the set point gradually, as shown by AI1/OUT.

When communications are re-established, the input process variables to PIDPLUS/PID2 reflect the true measurement provided by AI1/AI2, respectively. For PID2, AI2/OUT is very low compared to the set point, which causes a sharp spike in AO2 by the derivative part of PID2 at the moment communications are re-established. For PIDPLUS, AI1/OUT is close to the set point, and that small deviation is further evened out by the divisor used in the derivative part of the new algorithm. Thus both AI1 and AO1 transit to their steady states smoothly.

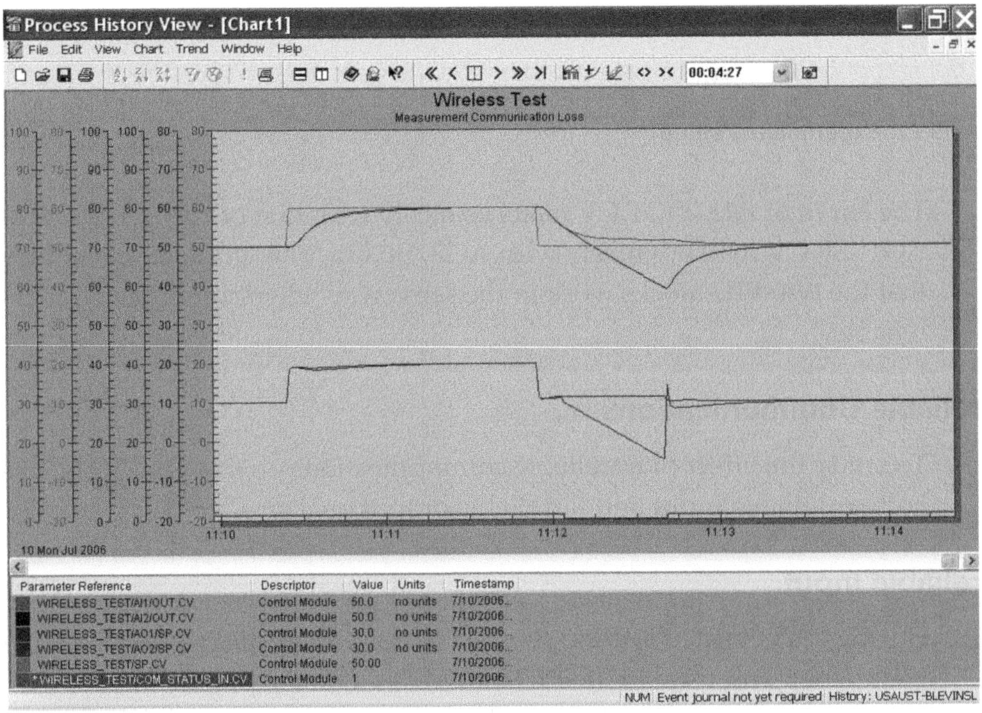

Figure 9-29. Lost inputs coupled with a set point change

The different behaviors of the two PID bocks are further proved by the performance data. In a duration of 121 seconds, the IAE for PIDPLUS is 169, while that for PID2 is 372.

Not shown are the effects of using a process disturbance instead of set-point change. The results were very close to what was observed for the set-point change [3].

Unreliable Output

In this case, the behaviors of the two PID blocks are examined with unreliable output communications. To test out the PID enhancements, the set point is changed from 50 to 60 when the communication is reliable. Then, after the processes settle at the set point, the set point is changed back to 50, and the output channels are cut off. Figure 9-29 shows the resulting curves.

When communication is lost, the outputs of PIDPLUS/PID2 are equal. Since the two controlled processes are the same, the values of AI1.OUT and AI2.OUT follow the same curve. When communication is reestablished, the input errors for PIDPLUS and PID2 are the same. However, the divisor in the derivative part of PIDPLUS is much bigger than that in PID2, which explains the sharper spike in the curve for AO2/OUT. During the transition period, the IAE for PIDPLUS is 190, while that for PID2 is 196.

9-6. Installation of a Wireless Network

Installing a wireless network includes installing wireless devices, a gateway (in some cases both a gateway and one or more access points), and a connection to a host or control system. Once the gateway, devices, and control system are configured the wireless mesh forms and communications begin.

Devices

As has been discussed, before devices can join the wireless mesh network they must be given a Network ID and a join key. The device uses the Network ID to determine which advertisement message to join through (there could be more than one wireless network within range). Once the device has heard one or more neighbors, it uses the join key to send a join request to the network manager. The network manager in turn authenticates the device and incorporates the device into the network.

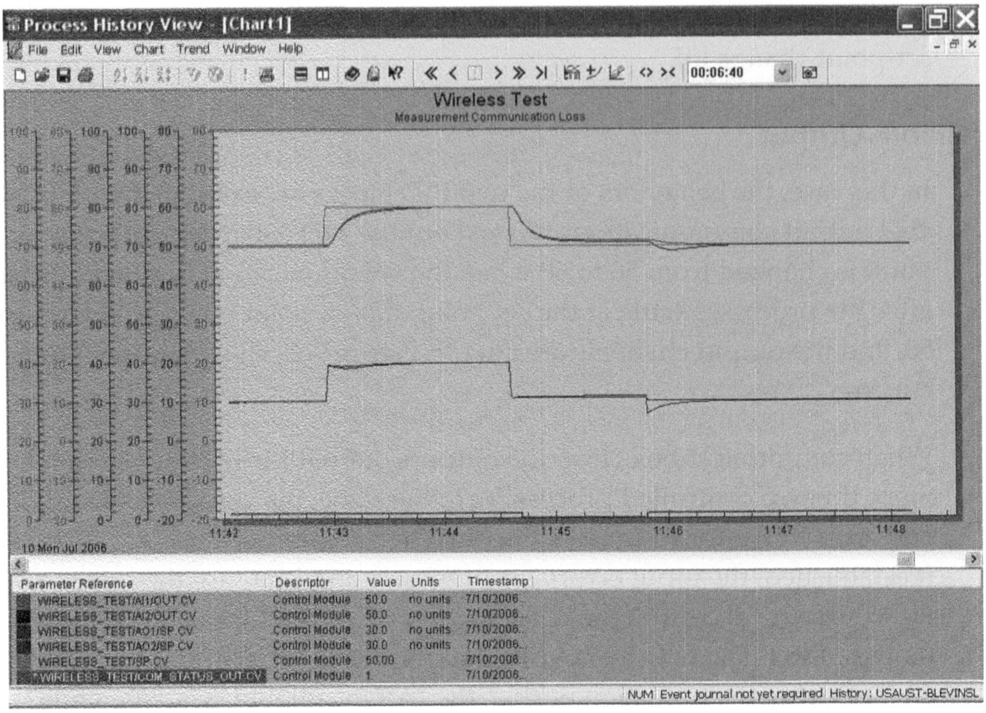

Figure 9-29. Lost inputs coupled with a set-point change

Gateways

The gateway provides the connection between host applications and control systems and the devices that make up the wireless network. The gateway is responsible for collecting and maintaining cached response messages from all devices in the network. These cached response messages originate from burst mode commands, event notifications, and commonly communicated HART commands. These cached response messages are returned as responses to host-based application requests. The gateway has additional built-in functionality to support adapters [5] that allow transparent access to the devices connected to an adapter.

In many cases the gateway connects to the mesh network through multiple access points (Figure 9-19). When multiple access points are in use the network manager will schedule communication traffic through all of them. If one of these access points fails then the network manager will adjust the schedule by spreading traffic across the remaining access points.

An example of a gateway is the Rosemount 1420 Wireless Gateway (Figure 9-30) [21][27].

Integration with Control Systems

Figure 9-30 shows how one or more wireless gateways are connected to the control system.

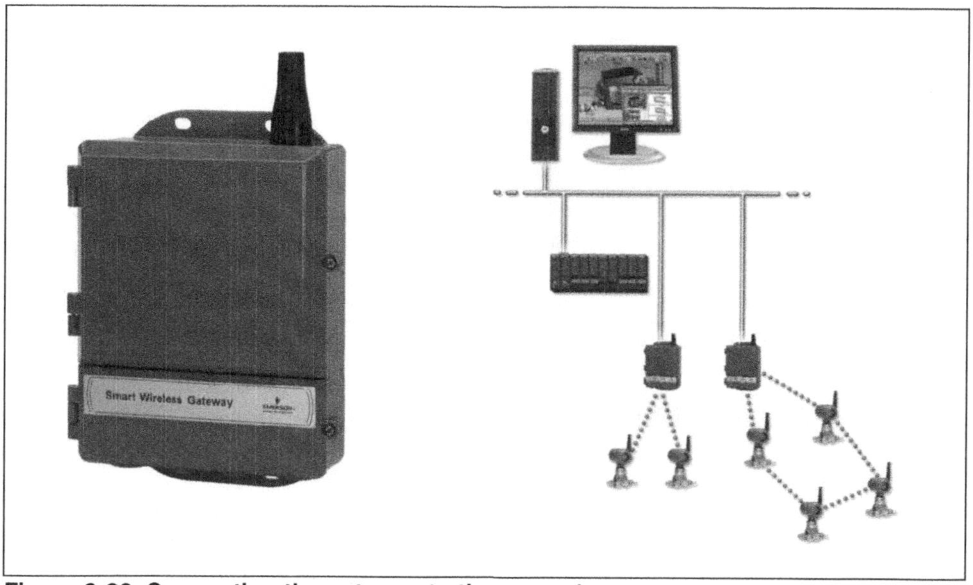

Figure 9-30. Connecting the gateway to the control system

In many situations the gateway connects to the control system through native interfaces and shows up as a part of the overall configuration system (Figure 9-31). In other cases, industry standard mechanisms (e.g., Modbus and OPC) can be used.

Exercises

1. What are key features of smart devices, and how do these features improve operations in the plant?
2. What are key advantages of wireless networks?
3. What is a mesh topology, and why is it used in industrial settings?
4. In wireless device networks, what are several key features required for improved measurement and control? How do these differ from wired networks?

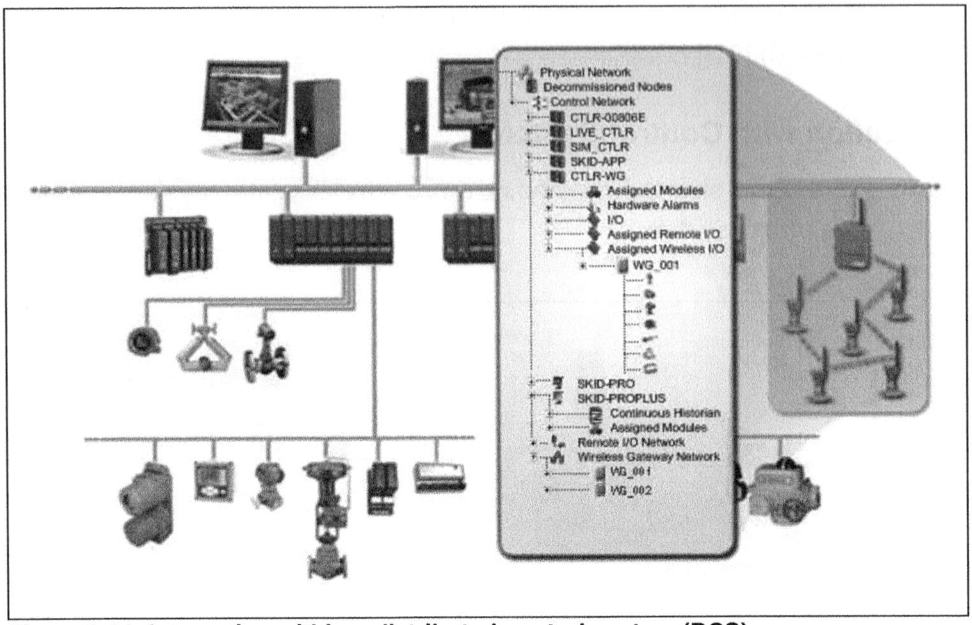

Figure 9-31. Integration within a distributed control system (DCS)

5. What is TDMA, and how are communications scheduled in a WirelessHART network?
6. How is high reliability achieved in a wireless network?
7. How are communications secured in a WirelessHART network?
8. A side effect of equipment trip is a sudden increase in the number of communications. How does a wireless network handle this surge in communications?
9. How can oversampling be reduced in a wireless network?
10. How do wired communication systems address variation in communication times? How do control strategies handle missed communications?
11. How can wireless communication systems address this same variation in communication times?
12. What is involved in installing a wireless network? What must devices be given in order to join a wireless network?

References

1. Nixon, M., Chen, D., Blevins, T., and Mok, A. K. "Meeting Control Performance over a Wireless Mesh Network." Presented at IEEE CASE Technical Conference, 2008.

2. McMillan, G. K. "Is Wireless Process Control Ready for Prime Time?" *Control*, May 2009.

3. Song, J., Mok, A. K., Chen, D., Nixon, M., Blevins, T., and Wojsznis, W. "Improving PID Control with Unreliable Communications." Presented at ISA EXPO Technical Conference, 2006.

4. Blevins, T., McMillan, G. K., Wojsznis, W., and Brown, M. "Advanced Control Unleashed: Plant Performance Management for Optimum Benefit." Research Triangle Park: ISA, 2002.

5. HART Communication Foundation (HCF) WirelessHART Standard, Sep. 2007.

6. IEEE 802.15.4 Standard, 2006.

7. Song, J., Han, S., Mok, A. K., Chen, D., Lucas, M., and Nixon, M. "A Study of Process Data Transmission Scheduling in Wireless Mesh Networks." ISA EXPO Technical Conference, Oct. 2007.

8. Song, J., Han, S., Mok, A. K., Chen, D., Lucas, M., Nixon, M, and Pratt, W. "WirelessHART: Applying Wireless Technology in Real-Time Industrial Process Control." 14th IEEE Real-Time and Embedded Technology and Applications Symposium, 2008.

9. Dworkin, M. "Recommendation for Block Cipher Modes of Operation: The CCM Mode for Authentication and Confidentiality." *NIST Special Publication 800-38C*, May 2004.

10. *FIPS Publication 197, Advanced Encryption Standard (AES)*. U.S. DoC/NIST, Nov. 26, 2001.

11. Diffie, W. and Hellman, M. "Privacy and Authentication: An Introduction to Cryptography." *Proceedings of the IEEE*, Vol. 67 No. 3, Mar. 1979, pp. 397-427.

12. ISO 7498-1 Information Processing Systems – OSI Reference Model – The Basic Model.

13. Halsall, F. *Data Communications, Computer Networks and Open System, Third Edition*. Addison Wesley, 1992.

14. Chen, D., Nixon, M., Aneweer, T., Shepard, R., Blevins, T., McMillan, G. K., and Mok, A. K. "Similarity-based Traffic Reduction to Increase Battery Life in a Wireless Process Control Network." Chicago: ISA EXPO 2005.

15. Foundation Fieldbus standard, http://www.fieldbus.org/.

16. International Society of Automation, http://www.isa.org/.
17. Profibus standard, http://www.profibus.org/.
18. Wireless Industrial Networking Alliance, http://www.wina.org/.
19. ZigBee Alliance, http://www.zigbee.org/.
20. Chipcon Products, http://www.chipcon.com/.
21. DeltaV digital control system, http://www.easydeltav.com/.
22. Caro, D. "Wireless Networks for Industrial Automation." Research Triangle Park: ISA, 2004.
23. IEC 61804-3 standard, Electronic Device Description Language (EDDL).
24. Control Information from John Shaw, http://learncontrol.com/pid.
25. "PID Controller." http://en.wikipedia.org/wiki/PID_controller.
26. http://www.emersonprocess.com/rosemount/document/pds/00813-0100-4420.pdf.
27. "Opportunities for Smart Wireless pH, Conductivity Measurements," http://www.isa.org/InTechTemplate.cfm?Section=Archives4&template=/ContentManagement/ContentDisplay.cfm&ContentID=80886.
28. "DeltaV Version 11 PID Enhancements for Wireless," August 2010, http://www.EmersonProcessXperts.com/papers/DeltaV-Version11-PID-Enhancements-for-Wireless.pdf.
29. McMillan, Greg, "Wireless – Overcoming Challenges of PID Control & Analyzer Applications," *InTech*, July-August 2010.
30. "Wireless PID Benefits Extend to Abnormal Situations, Analyzers, and Valves," August 5, 2010, http://modelingandcontrol.com.

Appendix A
Suggested Readings and Study Materials

Blevins, Terrence L., Gregory K. McMillan, Willy K. Wojsznis, and Michael W. Brown. *Advanced Control Unleashed: Plant Performance Management for Optimum Benefit*, ISA, 2003.

Magison, E. C., *Temperature Measurement in Industry*, ISA, 1990.

McMillan, Gregory K. "Wireless - Overcoming challenges of PID control & analyzer applications," *InTech Magazine*, July-August, 2010.

Murrill, P. W., *Fundamentals of Process Control Theory*, 3rd edition, ISA, 2000.

Spitzer, D. W., *Advanced Regulatory Control: Applications and Techniques*, Momentum Press, 2010.

Appendix B
Solutions to Exercises

Chapter 1 – Temperature Measurement

1-1. The RTD offers an order of magnitude better sensitivity, drift, and linearity plus online calibration and interchangeability of sensors.

1-2. Thermocouples offer a larger temperature range, smaller diameter, and an ability to withstand greater vibration. The often state advantage of faster sensor speed of response does not come into play unless the bare element is exposed to the process (a rare case).

1-3. A Thermistor is used for narrow range high sensitivity bare element applications such as internal mechanism or electronics temperature measurements.

1-4. Optical pyrometers are used in rotating equipment, moving solids, and furnaces where contact of a thermowell is not practical or sensor life and drift is unacceptable due to high temperatures and chemical attack (corrosion, oxidation, and reduction).

1-5. Changes in target emittance, intervening gases and particles, and widow conditions.

1-6. A 2-color optical pyrometer compensates for changes in emittance that are equal for both wavelengths.

1-7. The lead wire error in a 4-wire RTD installation is significantly less than in a 3-wire RTD installation when the difference in lead wire resistance is significant due to long runs and an appreciable tolerance in lead wire resistance (e.g., 10%).

1-8. Dual spring-loaded isolated and grounded thermocouple elements provide the best speed of response and reliability.

1-9. A tapered or swaged (stepped) thermowell offers a lower conduction error and vibration failure.

1-10. Transmitters eliminate lead wire errors and provide a more robust signal, match of calibration to sensor, narrowed span, and better maintenance (visibility and diagnostics).

Chapter 2 – Measurement Error

2-1. A. Thermal resistance between the process and the thermowell interior tip. This resistance is a function of the heat transfer coefficient of the process, the cross-sectional area of the thermowell tip, the thermowell tip thickness and thermal conductivity of the thermowell tip materials. For a reduction in heat conduction error this resistance should be small.

B. Thermal resistance between the thermowell interior tip and the process exterior connection. This resistance is a function of the thermowell wall length, the thermal conductivity of the thermowell wall and the heat transfer area of the thermowell wall. For a reduction in heat conduction error this resistance should be large minimizing the affect of the outside ambient temperature.

2-2.

$$A_w = \text{heat transfer area of thermowell wall}$$

$$= 3.14(r_o^2 - r_i^2)\ ft^2$$

$$= 3.14(((.25/12/2) + (.25/12/2))^2 - (.25/2/12)^2)$$

$$= .004192\ ft^2$$

$$R_w = \text{thermal resistance along the wall}$$

$$= (x_w/k_w) * (1/A_w)$$

$$k_w = 9\ Btu/hr * ft * {}^\circ F \text{ from Appendix D}$$

$$R_w = (18/12/9) * (1/.004192)$$

$$= 39.75^\circ F * hr/Btu$$

2-3.

p	k	m	h	D = r x 2	Cc	L
=(B2*C2-D2)/(B2*C2+D2)	6	=(4*D2/B2/E2)^0.5	10	0.03	=((1+A2)*EXP(-C2*G2))/(1+A2*EXP(-2*C2*G2))	0.75
=(B3*C3-D3)/(B3*C3+D3)	6	=(4*D3/B3/E3)^0.5	10	=E2+0.02	=((1+A3)*EXP(-C3*G3))/(1+A3*EXP(-2*C3*G3))	0.75
=(B4*C4-D4)/(B4*C4+D4)	6	=(4*D4/B4/E4)^0.5	10	=E3+0.02	=((1+A4)*EXP(-C4*G4))/(1+A4*EXP(-2*C4*G4))	0.75
=(B4*C5-D5)/(B5*C5+D5)	6	=(4*D5/B5/E5)^0.5	10	=E4+0.02	=((1+A5)*EXP(-C5*G5))/(1+A5*EXP(-2*C5*G5))	0.75
=(B6*C6-D6)/(B6*C6+D6)	6	=(4*D6/B6/E6)^0.5	10	=E5+0.02	=((1+A6)*EXP(-C6*G6))/(1+A6*EXP(-2*C6*G6))	0.75
=(B7*C7-D7)/(B7*C7+D7)	6	=(4*D7/B7/E7)^0.5	10	=E6+0.02	=((1+A7)*EXP(-C7*G7))/(1+A7*EXP(-2*C7*G7))	0.75
=(B8*C8-D8)/(B8*C8+D8)	6	=(4*D8/B8/E8)^0.5	10	=E7+0.02	=((1+A8)*EXP(-C8*G8))/(1+A8*EXP(-2*C8*G8))	0.75
=(B9*C9-D9)/(B9*C9+D9)	6	=(4*D9/B9/E9)^0.5	10	=E8+0.02	=((1+A9)*EXP(-C9*G9))/(1+A9*EXP(-2*C9*G9))	0.75

=(B10*C10-D10)/(B10*C10+D10)	6	=(4*D10/B10/E10)^0.5	10	=E9+0.02	=((1+A10)*EXP(-C10*G10))/(1+A10*EXP(-2*C10*G10))	0.75
=(B11*C11-D11)/(B11*C11+D11)	6	=(4*D11/B11/E11)^0.5	10	=E10+0.02	=((1+A11)*EXP(-C11*G11))/(1+A11*EXP(-2*C11*G11))	0.75
=(B12*C12-D12)/(B12*C12+D12)	6	=(4*D12/B12/E12)^0.5	10	=E11+0.02	=((1+A12)*EXP(-C12*G12))/(1+A12*EXP(-2*C12*G12))	0.75

p	k	m	h	D = r x 2	Ce	L
0.798879192	6	14.9071198	10	0.03	2.50866E-05	0.75
0.747735033	6	11.5470054	10	0.05	0.000302953	0.75
0.708258201	6	9.75900073	10	0.07	0.001131992	0.75
0.676534198	6	8.60662966	10	0.09	0.00263508	0.75
0.64732812	6	7.78498944	10	0.11	0.004797823	0.75
0.622405637	6	7.16114874	10	0.13	0.007544285	0.75
0.6	6	6.66666667	10	0.15	0.010780422	0.75
0.57959751	6	8.26224291	10	0.17	0.014413927	0.75
0.560834642	6	5.92348878	10	0.19	0.018361972	0.75
0.543443311	6	5.6343617	10	0.21	0.22553354	0.75
0.527219332	6	5.38381902	10	0.23	0.026928244	0.75
0.512003098	6	5.16397779	10	0.25	0.03143701	0.75

D = Diameter
= .09 feet
= 1.08 inches

2-4. The best installation is a tapered or swaged thermowell installed in a pipe elbow. This installation allows the probe to be centered in the stream giving the most accurate indication of the bulk temperature. the tapered end reduces the heat conduction error while the larger end allows longer insertion lengths and more mechanical stability.

2-5. Radiation errors may be reduced by:

- Polishing the exterior of the high temperature probe surface to reduce its emissivity.
- Increasing the velocity of the material passing a thermowell to increase the convective heat transfer.
- Increasing the length of the probe and decreasing the diameter of the probe to reduce conduction errors.
- Adding one or more radiation shields between the probe and the lower temperature surfaces.

2-6. $$E_2 = \frac{(G * E * r) * (T_o^4 - T_d^4)}{(h * L)}$$

To = 450°F = 910°R

Td = 1000°F = 1460°R

r = .5 inch = 0.416 ft

L = 10 inch = .833 ft

h = 25 Btu/hr * ft^2 * °F from Appendix H, Table H-1

G = $.1783 \times 10^{-8}$ Btu/hr * ft^2 * $°R^4$

E = .262 from Appendix E

Note: $°R^4$ terms cancel out giving us an answer in °F

$$E_2 = \frac{0.1783 \times 10^{-8} * 0.252 * 0.0416 * (910^4 - 1460^4)}{25 * 0.833}$$

$$= 1783 \times 10^{-8} * 0.252 * 0.0416 * (910^4 - 1460^4)25 * \dot{0}.833$$

$$= 3.46°F$$

2-7.

T = $.03t^2 + 10t + 22$

dT/dt = .06t + 10

$dT/dt|_{12}$ = .06 * 12 + 10 = 10.72°F/min

E_3 = (dT/dt) * TCm = (10.72 * 45)/60 = 8.04°F

2-8.

E_4 = 0.5 * (r – 1.0) * (C_p/C_v + 1) * M^2 * Tg

M = 600/1211 = .465

T_g = 150 + 460 = 610°R

r = .75 for perpendicular flow

C_p/C_v = 1.4 for air

E_4 = 0.5 * (.75 – 1.0) * (1.4 + 1) * $(.465)^2$ * 610

= –39.56°R = –420.44°F

Note: The answer is negative since the sensed temperature is less than the total temperature.

2-9.

- RTDs are an order of magnitude more repeatable than thermocouples.
- Thermistors are the most sensitive to temperature changes.
- Thermocouples have the highest temperature range extending up to 3100°F.

2-10.

- Reset has a small effect on controller output for properly tuned loops.
- Doubling the derivative setting has no effect on the controller output since the loop filter is also doubled as a result of the change.
- Doubling the controller gain results in doubling the controller output change due to A/D conversion.

2-11.

- In thermocouples high temperatures and temperature cycling results in decalibration because of changes in composition of the two thermoelements.
- Resistance temperature detectors decalibrate because of strain, moisture and grain growth at extreme temperatures. Sensor coils change length and diameter at high temperatures thus changing their resistance. Water molecules dissociate into hydrogen and oxygen at high temperatures and are absorbed by the platinum increasing the sensor resistance. Also grain growths in platinum occur at high temperatures reducing the mechanical stability of the RTD.

2-12. The twelve errors are:

1. Heat conduction error
2. Dynamic error
3. Radiation error
4. Velocity error
5. Electronic error
6. Sensor error or tolerance limit
7. Nonlinearity error
8. Decalibration error
9. Insulation error
10. Leadwire error
11. Head termination error
12. Instrument termination error

Errors 1-4 are a result of the installation into the process and the process conditions. Errors 6-9 are a function of the type of sensor and the way it is manufactured. Errors 5 and 10-12 result from the control system and the way the sensor is hooked up to the control system.

Chapter 3 – Basic Feedback Control

3-1. Unknowns, loads, and disturbances will cause process deviations unless corrected by feedback control as typically provide by a PID loop.

3-2. Lack of familiarity with such high gains and a concern about large abrupt movements of the control valve lead users to use much lower gains. The real limitation is often measurement noise that is amplified by high controller gains.

3-3. For a step change in setpoint, the proportional mode provides a step change in output from gain action that makes the approach to setpoint faster compared to the ramping gradual integral action.

3-4. Ideally, the rate time should be set equal to the second largest time constant in the loop that is normally a heat transfer lag or thermowell lag. The rate time must not be set equal to or larger than the reset time and must not cause valve dither form amplification of measurement noise.

3-5. The dynamic reset limit prevents the burst of oscillations for large upsets caused by the temperature controller output changing faster than a secondary jacket temperature loop, flow loop, or control valve can respond.

3-6. The inflection point takes a long time to appear due to the large process time constant of a large vessel's temperature. The process can be viewed and tuned as a "near integrator"

3-7. The gain setting should be increased so that the product of the controller gain and reset time is greater than the inverse of the integrating process gain.

3-8. Changes in process dynamics can provide an indication of possible degradation of process equipment and control valves and nonlinearities that be used to schedule changes in controller tuning.

3-9. A full throttle setpoint response (bang-bang control) provides the fastest approach by holding the controller output at its limit until the projected end point temperature in the next deadtime is equal to or just slightly short of the setpoint. The output is then set to its final resting value, held for one deadtime, and finally released for feedback correction.

Chapter 4 – Process Dynamics or Temperature Loop Analysis

4-1. For small actuators, the largest source of deadtime is the time it takes for the controller output to ramp through the deadband and resolution or sensitivity limit.

4-2. The integrated error is halved if the temperature controller gain is doubled.

4-3. The biggest measurement contributors to loop deadtime is poor A/D input card resolution and the thermowell lag from low velocities, poor fit of the sensor (internal air gap), and coatings.

4-4. A doubling of the reset time will double the period of the valve limit cycle, which will result in a doubled amplitude due to the proportionally less attenuation of the slower oscillation by the well mixed volume.

4-5. If the rate of change of batch temperature is used for the controlled variable, the response is self-regulating and the batch temperature profile slope can be controlled.

Chapter 5 – Exchangers

5-1. $$X_1 = \frac{F_C * C_C}{U * A} = \frac{3000 * 1.3}{8 * 300} = 1.625$$

$$X_2 = \frac{F_C}{Fh} = \frac{3000}{1900} = 1.58$$

Since both X1 and X2 ratios are below 3 and in the optimum range of 0.5 to 2, control will be possible.

5-2. A counter current setup, as shown in part A, offers the maximum heat exchange.

5-3. there are two reasons the output temperature of a heat exchanger is so hard to control. First, if the heat transfer area is too large or too small, the process gain will be too large or too small, respectively. If the process gain is too small, the control valve will have little effect on temperature. If the process gain is too large, a small change in control valve position causes large temperature swings. Second, if the flow is too small or large, the process gain will be too large or small, respectively. Again, in this case, the control valve will be unable to do its job.

5-4. $Kc' = \left(\frac{F'}{F}\right)^2 * Kc = \left(\frac{25}{50}\right)^2 * (0.3) = 0.075$

$Ti' = \left(\frac{F'}{F}\right) * Ti = \left(\frac{25}{50}\right) * (0.8) = 0.4$ min/repeat

$Td' = Td$

$Td' = 0.2$ min

5-5. Fc = Fh * Ch * (Thi – Tho) / {Cc * (Tci – Tco)}

Fc = (1900 * 1.1(300 – 200)) / {+1.0 * (50 – 90)}

Fc = –5225 lb/hr

Chapter 6 – Reactors

6-1. The most difficult disturbance to correct for while controlling reactor temperature is fouling or coating of the heat transfer surface. Mixing velocities are slow in reactors, and it is likely that a coating will form on the heat transfer surface. the coating will change the process gain, thus requiring a change in controller gain.

6-2. The most common way to control reactor temperature is to use cascade control. The outer loop controls reactor temperature and supplies a set point to the slave reactor coolant temperature loop. The coolant temperature loop compensates for coolant upsets before they affect the reactor temperature and makes the process gain more linear for the reactor loop.

6-3. The smallest recommended ratio for maximum to minimum stable gain settings for a reactor temperature loop is ten. Too much gain will result in large temperature excursion which could surpass the cooling capabilities of the reactor jacket. Too little gain could result in positive feedback and contribute to temperature excursions due to slow loop response.

6-4. Zero repeats per minute. Since reset action is being exhibited by the controller anytime there is a deviation from set point, it is possible for the controller to wind up and overshoot the set point, possibly going beyond the reactor's cooling capabilities.

6-5. If the coolant system is expected to be the main source of disturbances, then inlet coolant temperature control should be used because it responds quicker to coolant temperature changes. If reactor fouling or coating is expected to be the main source of disturbances, then outlet coolant temperature control should be used.

Chapter 7 – Columns

7-1. The first control scheme where the temperature controller manipulates the reflux flow is generally the best because it provides internal reflux control. It is particularly effective in compensating for a sudden change in outside temperature due to a rain storm or "blue northern" cold front which is often the largest and most disruptive upset.

7-2. The first control scheme will not work well due to poor distillate receiver level control if the distillate flow is small, the level measurement is noisy, or the receiver has an extremely large cross-sectional.

7-3. The addition of feed flow feedforward control where the distillate and/or steam flow are in ratio to the feed flow provides the biggest improvement in column performance. The column temperature controller corrects one of the ratios.

7-4. The dead time is so large due to the composition response of liquid volumes in series that a closed loop oscillation method will take too long (12 or more dead times). An open loop method that identifies the loop dead time and the initial ramp rate of the process variable during the next two dead times is the quickest (three dead times).

Chapter 8 – Vessels, Desuperheaters, Dyers, Kilns, Calciners, Crystallizers, Extruders, Chambers, and Rooms

8-1. Since it is a dead time dominate system like most other inline loops, the peak error for fast upsets such as flow or pressure will be large unless flow control, feedforward, or pressure control is used.

8-2. A filter on the moisture calculation not set to be larger than the reset setting of the temperature loop and larger than the residence time is the most frequent cause of control strategy failure because of positive feedback and inverse response.

8-3. The inability to make an accurate temperature measurement of solids in a high temperature rotating cylinder and the interaction of inlet and outlet temperature loops are the most common problems.

8-4. A control strategy, such as straight temperature control or fixed cooling, where the initial cooling rate is maximized, will cause fines and coating of the heat transfer surfaces.

8-5. Room environmental conditions can be better controlled by:

   a. A dew point loop to eliminate interactions between the temperature and humidity loops.

   b. Fast, small bare element sensors to reduce the loop dead time from the large sensor lag caused by low velocity air.

   c. Cascade control from room to duct controlled variables to improve the loop dynamics and compensate for utility upsets.

   d. Better air circulation flow and patterns to backmix the air.

Chapter 9 – Wireless

9-1. Smart devices incorporate features such as status on measured values, time stamps, event latching and confirmation, and block data transfer. The availability of additional variables, such as density and temperature for Coriolis mass flow meters, provides additional information useful for understanding changes in the process. Using these smart devices results in fewer trips into the plant, a much better controlled process, and significantly less downtime.

9-2. Wireless networks can result in reduced installation and commissioning costs, elimination of wiring noise and connection problems, and a reduction in the overall lifecycle.

9-3. In a mesh topology each device is connected to several other devices and all devices are capable of routing communications traffic. Mesh networks are self-healing: the network can still operate even when a device breaks down or a connection goes bad. As a result a mesh topology is very reliable.

9-4. Several key features for improved measurement and control with wireless networks are values reported with status, exception reporting, and timestamps. Status is important to the control strategy, exception reporting significantly reduces the number of communications, and timestamps can be used to measure latency and jitter. Although all of these same features are also important for wired communications, in wireless communications exception reporting improves battery life and timestamps can be used to adjust control calculations for variations.

9-5. TDMA stands for Time Division Multiple Access. TDMA communications are scheduled by a centralized network manager. The network manager is responsible for the creation of this schedule and the associated connections. It is also responsible for the distribution of this schedule to the individual network devices.

9-6. The wireless network achieves very high reliability through the use of several mechanisms: multiple paths to devices, multiple RF channels, and multiple communication tries. If improved reliability is required, more paths can be inserted by the network manager. As additional devices are added to the mesh (network) the strength of the overall mesh gets better and better. Additional devices improve path diversity. Additional access points, and devices in general, increase throughput, reduce latency, and can be used to route around areas of potential interference.

9-7. Communications are secured by ensuring that only devices approved to join the network join, that communications are not tampered with, and that bad guys cannot spoof the network through the use of replay attacks (a replay attack occurs when the same packets are resent on the same network). Communications are made reliable by retrying on different channels, retrying at different points in time, and communicating on different paths.

9-8. When there is contention for network resources, wireless networks use message priority determine which packets are sent first. Message priority helps mitigate situations where devices need to perform flow control. Priorities are set so that network management commands are the highest priority (this allows the network manager to temporarily increase communication resources, re-route, etc.), followed by real-time published messages, followed by alarms and alerts. In the case of alarms and alerts, since they are time-stamped at the device, no critical information is lost if an alarm or alert is delayed a few timeslots.

9-9. With wireless systems, because of the inherent time-synchronized nature of the mesh network, it is possible to reduce or eliminate this oversampling. Sampling can be scheduled to occur just before communications are scheduled.

9-10. Current control strategies address variation in communication by oversampling the I/O system. In standard process control systems, a missed I/O communication is considered an error. Usually, a control loop is configured with a maximum number of lost I/O communications, after which the loop declares failure and the values are set according to the fail-safe configuration.

9-11. Wireless communication systems can address variation in communications by modifying the calculation of the integral and derivative components of PID algorithms. The modification detects missed communications and compensates for the variation in communication times.

9-12. Installing a wireless network includes installing wireless devices, a gateway and one or more access points, and a connection to a host or control system. Once the gateway, devices, and control system are configured the wireless mesh forms and communications begin. Before devices can join the wireless mesh network they must be given a Network ID and a join key. The device uses the Network ID to determine which advertisement message to join through.

Appendix C
Unification of Controller Tuning Relationships

An intensive search is underway for a unified field theory that would bring together quantum and gravitational forces and provide the underlying truth behind physical laws. In this appendix we show that we are at least fortunate enough to have achieved a unification of Lambda, internal model control, the Ziegler-Nichols reaction curve, and ultimate oscillation tuning methods for bioprocess control. For the control of vessel temperature, concentration, and gas pressure, the controller tuning equations from diverse methods are reducible to a common form, in which the maximum controller gain is proportional to the time-constant-to-dead-time ratio (τ_1 / θ_o) and is inversely proportional to the open loop gain (K_o), commonly known as the process gain. This common form is easy to remember and provides insight into the relative effects of process dynamics on tuning and hence on loop performance. This appendix concludes with a derivation of the equation to predict the control error (integrated absolute error) in terms of the tuning settings from a PI controller's response to load disturbances.

Lambda tuning provides stable results for any Lambda value. Normally, Lambda is set large enough to provide the degree of slowness desired to reduce interaction and promote the coordination of loops. If Lambda is set much smaller than normally expected, as outlined in this section, then the result is the common form, which provides maximum disturbance rejection and minimum integrated absolute error. Thus, Lambda tuning has the advantage of enabling the user to achieve a variety of objectives by setting the degree of transfer of variability from the process output to process input.

The Lambda tuning equations for self-regulating processes are as follows:

$$\lambda = \lambda_f * \tau_1 \quad \text{(C-1)}$$

$$T_i = \tau_1 \quad \text{(C-2)}$$

$$K_c = \frac{T_i}{K_o * (\lambda + \theta_o)} \quad \text{(C-3)}$$

$$T_d = \tau_2 \quad \text{(C-4)}$$

$$K_o = K_{mv} * K_{pv} * K_{cv} \quad \text{(C-5)}$$

The Lambda factor is the ratio of closed loop time constant (Figure 3-3a) to the open-loop time, where the open-loop time constant is the largest time constant (τ_1). For maximum load rejection capability, a Lambda equal to the total loop dead time ($\lambda = \theta_o$) can be used, and the loop will still be stable if the dynamics are accurately known. For many temperature loops, this corresponds roughly to a 0.1 Lambda factor ($\lambda_f = 0.1$). If we substitute this small Lambda into Equation C-3, we end up with the simplified internal model control (SIMC) equation (see Equation C-6). This equation has recently been documented as providing the best load rejection (tightest control) for self-regulating processes [1].

Until recently, proportional band (PB) was predominantly used rather than controller gain. The equation for minimum proportional band is 100 percent divided by Equation C-6 since the proportional band is 100 percent divided by the controller gain (PB =100% / K_c). Proportional band is the percent change in the control error that will cause a 100 percent change in controller output from the proportional mode.

$$K_c = 0.5 * \frac{\tau_1}{K_o * \theta_o} \quad \text{(C-6)}$$

Equation C-6 has dominated the literature since the days of Ziegler and Nichols. The multiplier ranges from 0.4 to 0.8, and the exponent of the time-constant-to-dead-time ratio varies from 0.9 to 1.0. The differences in the multipliers or exponents are insignificant because in practice the user has backed off from the maximum gain, and the effect of errors or changes commonly seen in the identified process gain, time constant, and dead time is larger than the effect of the coefficients.

If we substitute the definition of a pseudo or "near" integrator gain per Equation C-7 into Equation C-6 we end up with the Equation C-8, which is the SIMC tuning that has been shown to provide the tightest control of integrating processes [1].

$$K_i = \frac{K_o}{\tau_1} \tag{C-7}$$

$$K_c = 0.5 * \frac{1}{K_i * \theta_o} \tag{C-8}$$

If we change nomenclature such that the integrating process gain is the reaction rate ($K_i = R$) and the dead time is the delay time ($\theta_o = L$), then we end up with the controller gain per the Ziegler-Nichols "reaction curve" method developed in the 1940s.

The "reaction curve" method is suitable for self-regulating processes that have large time constants and integrating processes. However, the documentation of the "reaction curve" method showed that the process was lined out ($CV_1/\Delta t = 0$) just before the change is made in the controller output. As a result, R could be computed from just the ramp rate of the process variable after the change. This may not be the case, especially for integrating processes, since the controller is in manual. Therefore, it is critical to take into account the change in ramp rates from "before" to "after" the change and to use percent rather than engineering units, as shown in Equation C-9. The "short cut" method presented in the book *Good Tuning: A Pocket Guide, 2nd Edition* provides a detailed procedure for using Equation C-9 and correcting the observed dead time for the effect of final element resolution or dead band so as to quickly estimate the controller tuning for slow processes [2].

$$K_i = \frac{CV_2/\Delta t - CV_1/\Delta t}{\Delta CO} \tag{C-9}$$

It may be difficult to accurately identify the second-largest time constant. Therefore, the internal model control computation of the derivative time as shown in Equation C-10 may be useful. For a first-order (single time constant) approximation of a concentration or temperature response, about half of the total loop dead time originates from the second-largest time constant, as shown in Equation C-11. Since in this case the time constant is much larger than the dead time, the dead-time term in the denominator of Equation C-10 becomes just twice the largest time constant ($2*\tau_1$). If you then cancel out the time constant in the numerator and denominator, you end up with the derivative time being approximately equal to half the dead time. Equation C-10 coupled with Equation C-11 is reducible to Equation C-4.

$$T_d = \frac{\tau_1 * \theta_o}{2 * \tau_1 + \theta_o} \tag{C-10}$$

For interactive lags found in temperature, concentration, and gas pressure processes:

$$\tau_2 = 0.5 * \theta_o \tag{C-11}$$

The Lambda tuning equations for integrating processes are as follows:

$$\lambda = \lambda_f \mathbin{/} K_i \tag{C-12}$$

$$T_i = 2 * \lambda + \theta_o \tag{C-13}$$

$$K_c = \frac{T_i}{K_i * (\lambda + \theta_o)^2} \tag{C-14}$$

The Lambda factor is the ratio of closed loop arrest time to open-loop arrest time where the open loop arrest time is simply the inverse of the integrating process gain ($1/K_i$). For maximum load rejection capability, a Lambda equal to the total loop dead time ($\lambda = \theta_o$) can again be used, and the loop will still be stable if the dynamics are accurately known. In this case, Equation C-14 reduces to Equation C-8, but with a multiplier of 0.75 instead of 0.5. The higher multiplier is insignificant since it is rarely desirable, and dynamics are seldom known accurately enough to take advantage of this increase in the controller gain.

For many processes that have a true or near-to-true integrating response, the controller gains computed by Equation C-8 are much higher than desired or needed. This is particularly the case for columns, reactors, and vessels since the disturbances are so slow. However, a much lower controller gain can lead to nearly sustained oscillations with a very long period. To prevent this from occurring, Equation C-15, which is developed from the transfer function for the closed loop response of an integrating process, can be used to ensure that the response is over damped.

The integral time needed to ensure an over-damped response in integrating processes is as follows:

$$T_i > \frac{4}{K_i * K_c} \tag{C-15}$$

The product of the controller gain and the process gain is much larger than one for temperature and concentration loops on well-mixed volumes because the time constant is so large, which means you can cancel out the product of gains in the numerator. If you then use Equation C-7 to get an equivalent process integrating gain, then Equation C-16 is reducible to Equation C-15.

The integral time needed to ensure an over-damped response in self-regulating processes is as follows:

$$T_i > \frac{4 * (K_o * K_c * \tau_1)}{(1 + K_o * K_c)^2} \tag{C-16}$$

Ziegler and Nichols developed controller tuning equations based on field measurements of the ultimate gain and ultimate period. For a manual tuning test, the derivative time is set to zero, and the integral time is set at least ten times larger than normal so most of the controller response is from the proportional mode. The controller gain is then increased to create equal sustained oscillations. The controller gain at this point is the ultimate gain, and the oscillation period is the ultimate period. In industry, the gain is only increased until decaying oscillations first appear to reduce the disruption to the process. Auto tuners and adaptive controllers make this manual controller tuning unnecessary. The "relay method" is extensively employed by "on-demand" auto tuners to automatically compute the ultimate period and gain by switching the controller output when it crosses and departs from a noise band centered on the set point [2] [3].

The ultimate gain for self-regulating processes using the amplitude ratio is [3]:

$$K_u = \frac{[1 + (\tau_1 * 2 * \pi / T_u)^2]^{0.5}}{K_o} \tag{C-17}$$

For the Ziegler-Nichols ultimate oscillation method, the controller gain is simply a fraction of the ultimate gain, and the integral (reset) time is a fraction of the ultimate period, as the following equation shows for a PI controller:

$$K_c = 0.4 * K_u \tag{C-18}$$

$$T_i = 0.8 * T_u \tag{C-19}$$

If you also take into account that the squared expression in the numerator of Equation C-17 is much larger than one, you end up with Equation C-20. For most temperature and composition loops on columns, reactors, and vessels, the ultimate period is approximately four times the dead time ($T_u = 4 * \theta_o$). If you substitute this relationship into C-20, you end up with Equation C-21. After multiplying numerical factors, Equation C-21 becomes Equation C-22 (Equation C-6 with a slightly larger multiplier).

$$K_c = 0.4 * \frac{(\tau_1 * 2 * \pi)/T_u}{K_o} \tag{C-20}$$

$$K_c = 0.4 * \frac{(\tau_1 * 2 * \pi)/(4 * \theta_o)}{K_o} \tag{C-21}$$

$$K_c = 0.6 * \frac{\tau_1}{K_o * \theta_o} \tag{C-22}$$

If the ultimate period is about four times the dead time ($T_u = 4 * \theta_o$), then the integral time ends up as about three times the dead time, per Equation 3-1s for the ultimate oscillation method. This is the same result you get according to Equation C-13 for the Lambda tuning method when Lambda is reduced to equal the dead time. This reset time is generally considered to be too fast. The SIMC method states that while four times the dead time provides the best performance and that an increase to eight times the dead time provides better robustness. If four times the dead time is used, then Equations C-4 and C-11 result in a reset time that is eight times the rate time setting. Though most of the literature shows the rate time as being equal to one-quarter the reset time, in practice a rate time that is one-eighth to one-tenth the reset time provides a smoother response.

Equation 4-1 in chapter 4 for the integrated absolute error (*IAE*) can be derived from a PI controller's response to a load upset. The module execution time (Δt) is added to the reset or integral time (T_i) to show the effect of how the integral mode is implemented in some digital controllers. An integral time of zero ends up as a minimum integral time equal to the execution time so there is not a zero in the denominator of Equation C-23. For analog controllers, the execution time is effectively zero [6].

$$\Delta CO = K_c * \Delta E_t + [K_c/(T_i + \Delta t)] * \text{Integral}\ (E_t * \Delta t) \qquad \text{(C-23)}$$

The errors before the disturbance and after the controller has completely compensated for the disturbance are zero ($\Delta E_t = 0$). Therefore, the long-term effect of the proportional mode, which is first term in Equation C-23, is zero. Equation C-23 reduces to Equation C-24 [5].

$$\Delta CO = [(K_c/(T_i + \Delta t)] * \text{Integral}\ (E_t * \Delta t) \qquad \text{(C-24)}$$

For an over-damped response:

$$IAE = \text{Integral}\ (E_t * \Delta t) \qquad \text{(C-25)}$$

The open loop error is the peak error for a step disturbance in the case where the controller is in manual (loop is open). The open loop error (E_o) is the open loop gain (K_o) times the shift in controller output (ΔCO) required to compensate for the disturbance when the controller is in auto (loop is closed).

$$E_o = K_o * \Delta CO \qquad \text{(C-26)}$$

Equation C-24 solved for the *IAE* defined in Equation C-25 and the open loop error defined in Equation C-26 becomes Equation C-27. If you ignore the effect of module execution time (Δt) on the integral mode, Equation 4-2a in chapter 4 is Equation C-27 for an over-damped response because the integrated absolute error (*IAE*) is the same as the integrated error (E_i). Even for a slightly oscillatory response, the approximation has proven to be close enough [4].

$$IAE = \frac{1}{(K_o * K_c)} * (T_i + \Delta t) * E_o \qquad \text{(C-27)}$$

For the control of vessel temperature, concentration, and pressure, we can use Equation C-6 for the maximum controller gain and four times the dead time for the minimum reset time to express the minimum integrated absolute error in terms of the process dynamics. The resulting Equation

C-28 shows that the minimum integrated absolute error is proportional to the dead time squared for tuning settings that give the tightest control. Note that the open loop gain has cancelled out. This equation can be independently derived by multiplying the peak error for a step disturbance by the dead time [3] [6].

$$IAE = 2 * \frac{\theta_o}{\tau_1} * (4 * \theta_o + \Delta t) * E_o \qquad \text{(C-28)}$$

In practice, controllers are not tuned this aggressively. Often the reset time is set equal to the time constant and a Lambda factor of 1.0 is used, which corresponds to a controller gain that is about ten times smaller than the maximum controller gain for a reactor's primary loops (e.g., composition, pressure, and temperature).

If the disturbance is not a step change, the integrated absolute error will be smaller. The effect of a slow disturbance can be approximated by adding the disturbance time constant to the open loop time constant (τ_1) in the denominator of Equation C-28.

An increase in the module execution time shows up as an increase in the loop dead time for unmeasured disturbances. If the disturbance arrives immediately after the process variable is read as an input to the module, the additional dead time is about equal to the module execution time. If the disturbance arrives immediately before the process variable is read, the additional dead time is nearly zero. On the average, the additional dead time can be approximated as 50 percent of the module execution time. Simulations that create a disturbance that is coincident with the controller execution will not show much of an effect of execution time on performance. This scenario misleads users into thinking that the execution time of model predictive control is not important for load rejection. For chromatographs where the result is only available for transmission after the processing and analysis cycle, the additional dead time is 150 percent of the analyzer cycle time [2] [3] [5].

Equation C-28 shows the effect of the largest time constant, loop dead time, and module execution time on absolute integrated error if the controller is always retuned for maximum performance. A detuned controller may not do much better than a tightly tuned controller for a larger loop dead time or module execution time [5]. Thus, the value of reducing these delay times depends on the controller gain used in practice. For example, the controller gain is simply the inverse of the open loop gain for a Lambda factor of one in a loop with a dead time much smaller than the time constant. In other words, Equation C-3 reduces to Equation C-29.

$$K_c = \frac{1}{K_o} \qquad \text{(C-29)}$$

If you substitute Equation C-29 into Equation C-6 and solve for the dead time, you end up with Equation C-30. This shows a Lambda factor of one on a primary reactor loop, which implies a dead time that is about ½ of the time constant. The integrated absolute error for this case will not appreciably increase until the dead time is about ½ of the time constant. Thus, time and money spend on reducing the dead time or module execution time below this implied dead time has little value unless the controller is retuned [5].

$$\theta_o = 0.5 * \tau_1 \qquad \text{(C-30)}$$

Nomenclature

ΔCO = shift in controller output to compensate for disturbance (%)

E_i = integrated error (% seconds)

E_o = open loop error for a step disturbance (%)

E_t = error between SP and PV during the disturbance

IAE = integrated absolute error from the disturbance (% seconds)

K_c = controller gain (dimensionless)

K_i = integrating gain (%/sec/% or 1/sec)

K_o = open loop gain (dimensionless)

K_u = ultimate gain (dimensionless)

λ = Lambda (closed loop time constant or arrest time) (sec)

λ_f = Lambda factor (ratio of closed to open loop time constant or arrest time) (dimensionless)

Δt = module execution time (sec)

θ_o = total loop dead time (sec)

τ_1 = largest open loop time constant (sec)

τ_2 = second largest open loop time constant (sec)

T_i = integral (reset) time setting (sec/repeat)

T_d = derivative (rate) time setting (sec)

T_u = ultimate period (sec)

References

1. Skogestad, Sigurd, "Simple Rules for Model Reduction and PID Controller Tuning", *Journal of Process Control*, 13, (2003): 291-309.

2. McMillan, Gregory K., *Good Tuning: A Pocket Guide, 2d ed.*, ISA, 2005.

3. Blevins, Terrence L., McMillan, Gregory K., Wojsznis, Willy K., and Brown, Michael W., *Advanced Control Unleashed: Plant Performance Management for Optimum Benefits*, ISA, 2003.

4. Shinskey, F. G., *Feedback Controllers for the Process Industries*, McGraw-Hill, 1994.

5. McMillan, Gregory, K., *Tuning and Control Loop Performance, 3d ed.*, ISA, 1991.

6. Shinskey, F. G., "The Effect of Scan Period on Digital Control Loops", *InTech*, June 1993.

Appendix D
Physical Property Data

Tables nomenclature and units:

T = temperature (°F)

k = thermal conductivity of fluid (Btu/hr*ft*°F)

c = heat capacity of fluid (Btu/lb*°F)

d = density of fluid (lb/ft3)

Note that the reported values will differ by about 10% due to variations in test methods, temperature, and composition.

Table D-1. The thermowell material, coating, and fill have a significant effect on its thermal conductivity [1]

Material	k
Air	0.02
Carbon steel	30
Glass	0.62
Light heat transfer	0.08
Hastelloy B	6
Hastelloy C	5
316 stainless steel	9
Tantalum	31
Teflon	0.12

Table D-2. The thermowell material and fill have a significant effect on its thermal conductivity [1]

Material	k
Brass	60
Copper	220
Ethylene Glycol (liquid)	0.144
Hastelloy B or C	7
Monel	15
Nickel	45
Porcelain	1
304 stainless steel	13
316 stainless steel	10
Tantalum	32

Table D-3. The physical properties of fluids depend upon temperature [3]

Material	T	k	c	d
Metals:				
Aluminum	32	117	0.208	169
Brass	32	56	0.092	532
Bronze	32	15	0.082	540
Constantan	32	12	0.1	557
Copper	32	224	0.091	558
Iron, cast	32	33	0.11	474
Iron, pure	32	36	0.104	491
Nickel	32	35	0.103	555
Steel, mild	32	27	0.11	490
18-8 stainless steel	32	8	0.11	488
Insulators:				
Cork	86	0.025	0.04	10
Glass wool	100	0.031	—	1.5
Kaolin firebrick	392	0.110	—	19
85% magnesia	200	0.037	—	17
Rock wool	200	0.030	—	8
Rubber	32	0.087	0.48	75
Fire clay brick	392	0.580	0.20	144
Masonry	70	0.38	0.20	106
Zirconia	392	0.84	—	304
Fluids:				
Air	100	0.0154	0.240	0.071
Air	400	0.0212	0.245	0.046
Benzene	100	0.087	0.44	54.0
Light oil	100	0.076	.46	56.0
Glycerin	100	0.163	0.600	78.2
Steam	212	0.0145	0.451	0.0372
Steam	400	0.0200	0.462	0.0288
Water	50	0.332	1.00	62.4
Water	100	0.364	0.998	62.0

References

1. Kardos, P. W., "Response of Temperature Measuring Elements," *Chemical Engineering*, August 29, 1977, pg. 82.
2. Crawford, C. L., "Thermowell Heat Conduction Error Versus," Texas A&M Symposium, January, 1982, Appendix I, pg. 2.
3. Kreith, F., *Principles of Heat Transfer*, 2nd edition, International Textbook Company, 1965, pg. 376.

Appendix E
Emissivities

Table E-1. The emissivities of materials vary with surface condition and geometry and operating temperature [1]

Surface	Temperature, °F	Emissivity
Aluminum plate (unpolished)	78-1000	0.055-0.150
Oxidized steel	70-950	0.633-0.800
Rough cast iron	100-500	0.930-0.955
Oxide coated steel tube	900	0.750-0.795
301 stainless steel	450-1700	0.512-0.701
316 stainless steel	450-1900	0.262-0.662
310 stainless steel	460-1000	0.880-0.967
347 stainless steel	450-1600	0.500-0.655
304 stainless steel	900-980	0.600-0.740
Monel	450-1600	0.450-0.640
Alumina - fine grain surface	1800-2900	0.180-0.300
Alumina - coarse grain surface	1800-2900	0.410-0.500
Alumina - low iron	1800-2900	0.400-0.600
Alumina - medium iron	1800-2900	0.600-0.770
Alumina - high iron	1800-2900	0.680-0.790
Typical fireclay	1800	0.750

Reference

1. Reed, R. D., *Furnace Operations*, 3rd edition, Gulf Publishing Company, 1981, pg. 70.

Appendix F
First Principle Process Gains, Dead Times, and Time Constants

There are three types of processes; self-regulating, integrating, and runaway as shown in Figures F-1, F-2, and F-3, respectively. A self-regulating process will decelerate to a new steady state. An integrating process will continually ramp. A runaway process will accelerate until hitting a relief or interlock setting.

Over 90% of the processes are self-regulating. However, many of the continuous and fed-batch processes in the chemical industry with the greatest direct economic benefits behave and can be best treated as "near integrating" processes. The classic integrating process is a pure batch or level process. Less than 1% of the processes are runaway. When these exist, understanding the runaway response is critical in terms of safety and control because of the propensity to accelerate and reach a point of no return. Runaway responses are almost exclusively associated with highly exothermic reactors used in plastics and specialty chemical production.

This note develops the equations for process dynamics for back mixed volumes and plug flow volumes. The back mixed volumes section applies to volumes whenever an agitator pumping rate, an eductor or recirculation liquid flow rate, or gas evolution or sparge rate produces enough turbulence and back mixing to make the mixture more uniform in the axial besides the radial direction of the volume. Gas volumes generally have enough turbulence and a fast enough gas dispersion rates to be treated as a back mixed volume. The equations therefore hold relatively well for an evaporator and a single distillation stage due to turbulence from the vapor flow. The plug volume section applies to static mixers, pipelines, coil inlets, and jacket inlets where the turbulence from pipe fittings or internal mixing elements creates enough radial mixing to make the mixture uniform over the cross-section of the pipe or nozzle inlet but little axial mixing.

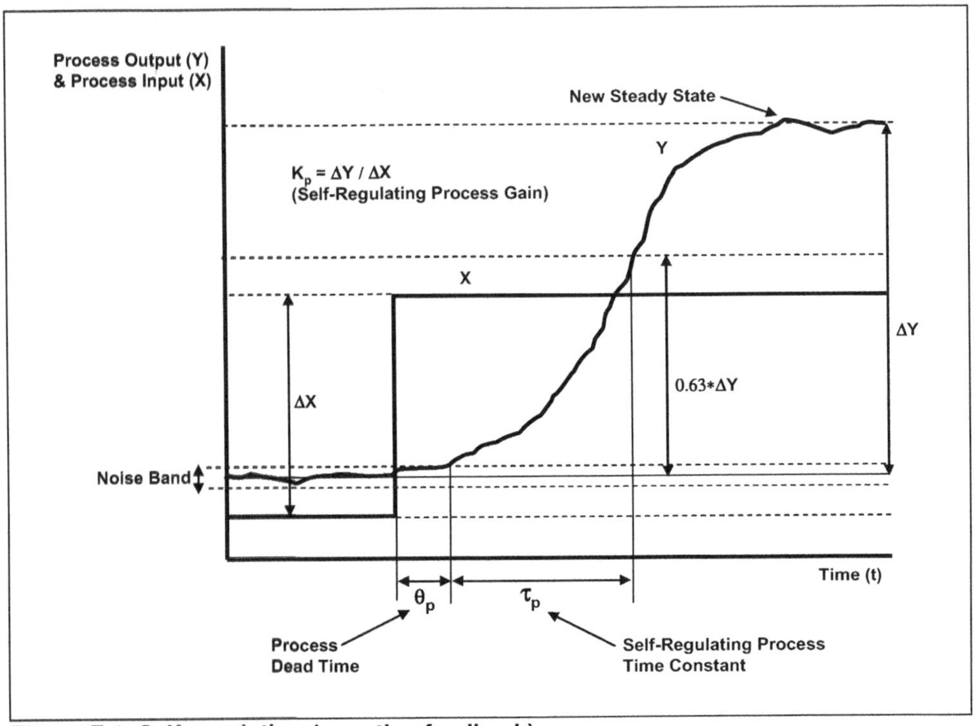

Figure F-1. Self-regulating (negative feedback) process

Back Mixed Volumes

For a back mixed volume, the process gains and time constants can be readily identified if the ordinary differential equations for the rate of accumulation of energy or material in the volume are set up so that the process output of interest (Y) is on right side with a unity coefficient. From this simple generic form we can identify the process time constant (τ_p) as the coefficient of derivative of the process output (dY/dt) and the process gain (K_p) as the coefficient of the process input (X). The process output (Y) and input (X) can be viewed as the controlled and manipulated variables, respectively. Many other terms can exist but these are not shown in the following equations. These missing terms can be categorized as disturbances.

If the sign of the unity coefficient of the process output on the right side is negative (Equation F-1a), the process has negative feed back. As the pro-

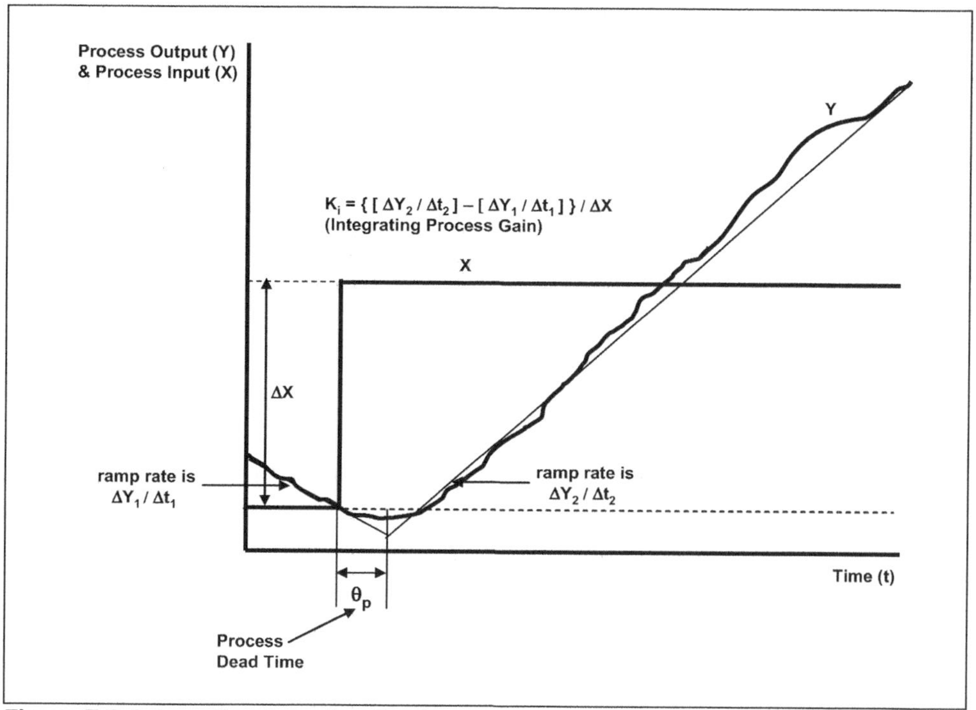

Figure F-2. Integrating (zero feedback) process

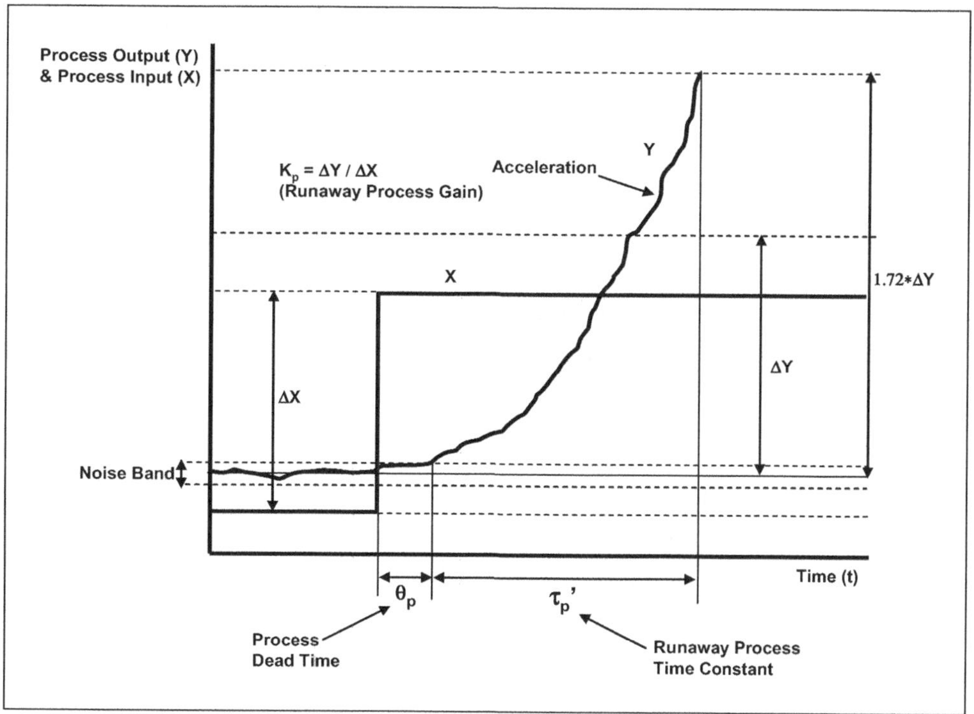

Figure F-3. Runaway (positive feedback) process

cess output changes, the negative feedback slows down and eventually halts the excursion of the process output at its new steady state when it balances out the effect of the process input and the disturbances.

$$\tau_p * dY / dt = K_p * X - Y \quad \text{(F-1a)}$$

The integration of this equation provides the time response of a change in the process output (ΔY) for a change in the process input (ΔX).

$$\Delta Y = K_p * (1 - e^{-t/\tau_p}) * \Delta X \quad \text{(F-1b)}$$

If the process output does not appear on the right side (Equation F-2a), there is no process feedback. As the process output changes, there is no feedback to slow it down or speed it up so it continues to ramp. There is no steady state. The ramping will only stop when X is zero or balances out the disturbances.

$$dY / dt = K_i * X \quad \text{(F-2a)}$$

$$\Delta Y = K_i * \Delta t * \Delta X \quad \text{(F-2b)}$$

Often in the more important loops for concentration, pressure, and temperature control of large volumes, the time constant in Equation F-1a is so large that the time to reach steady state is beyond the time frame of interest. Since these loops with small dead time to time constant ratios should be tuned with small Lambda factors (high controller gains) per Advanced Application Note 3, the controller only sees the first part of the excursion before the inflection point and deceleration by negative process feedback. In this case we have a "near Integrator" and Equation F-1a is best visualized as Equation F-2a with an integrator gain calculated per Equation F-2c.

$$K_i = K_p / \tau_p \quad \text{(F-2c)}$$

If the sign of the unity coefficient of the process output on the right side is positive (Equation F-3a), the process has positive feed back. As the process output changes, the positive feedback speeds up the excursion unless disturbances counteract the effect of the process input and output.

$$\tau_p * dY / dt = K_p * X + Y \quad \text{(F-3a)}$$

$$\Delta Y = K_p * (e^{t/\tau_p} - 1) * \Delta X \quad \text{(F-3b)}$$

Consider a mixed volume with a jacket and vapor space as shown in Figure F-4. There are liquid reactant feeds, gas feeds (sparged through the liquid and added directly to the vapor space), an outlet liquid flow, a vent gas flow, and a jacket coolant flow. There is normally multiple components interest. For example, consider liquid or gas acid and base reagent or reactant components (*a*, *b*) to produce primary and secondary liquid or gas products (*c*, *d*, *e*). Consider also there are typically water and nitrogen gas components (*w*, *n*).

The ordinary differential equation for the accumulation of liquid mass as shown in Equation F-4a includes inlet flows added directly to the liquid volume (ΣF_i), vapor flow rates from evaporation and vaporization (ΣF_v), and an outlet liquid flow rate (F_o). The liquid level depends upon density and cross-section area of the liquid. Equation F-4a can then be reformulated to Equation F-4b to include the process variable of interest, liquid level (L_l), in the derivative.

$$dM_o / dt = \Sigma F_i - \Sigma F_v - F_o \quad \text{(F-4a)}$$

$$d(\rho_o * A_o * L_o) / dt = \Sigma F_i - \Sigma F_v - F_o \quad \text{(F-4b)}$$

If we consider the density (ρ_o) to be a weak function of composition and therefore constant like the cross-sectional area (A_o) we can take these terms outside the derivative and divide through to get an equation for level (L_o) in the form of F-2a. Now it is clearly evident that the integrating process gain (K_i) for manipulation of flows in or flow out is simply the inverse of the product of the liquid density and cross-section area (Equation F-4d).

$$dL_o / dt = [1/(\rho_o * A_o)] * [\Sigma F_i - \Sigma F_v - F_o] \quad \text{(F-4c)}$$

$$K_i = 1/(\rho_o * A_o) \quad \text{(F-4d)}$$

The ordinary differential equation for the accumulation of gas mass as shown in Equation F-5a includes inlet flows added directly to the gas volume (ΣF_i), vapor flow rates from gas sparging, evolution, and vaporization (ΣF_v), and an exit gas flow rate (F_g). Equations of state such as the ideal gas law can be used to express this relationship for a given composition. Equation F-5a can then be reformulated to Equation F-5b to include the process variable of interest, gas pressure (P_g), in the derivative.

$$dM_g / dt = \Sigma F_i + \Sigma F_v - F_g \qquad \text{(F-5a)}$$

$$d[(P_g * V_g)/(R_g * T_g)]/dt = \Sigma F_i + \Sigma F_v - F_g \qquad \text{(F-5b)}$$

If we consider changes in the gas volume (V_g) and gas temperature (T_g) to be much slower than changes in the gas pressure (P_g) and therefore relatively constant during the integration step we can take these terms outside the derivative and divide through to get an equation for pressure in the form of Equation F-2a. Now it is clearly evident that the integrating process gain (K_i) for manipulation of flows in or flow out is simply the product of the universal gas coefficient (R_g) and the absolute gas temperature divided by the gas volume (Equation F-5d). This assumes a change in pressure does not significantly change the gas glow out of the volume, which is normally the case for a pressure drop across the vent valve that is large or critical.

$$dP_g/dt = [(R_g * T_g)/V_g] * [\Sigma F_i + \Sigma F_v - F_g] \qquad \text{(F-5c)}$$

$$K_i = [(R_g * T_g)/V_g] \qquad \text{(F-5d)}$$

The ordinary differential equation for the accumulation of energy as shown in Equation F-6a includes the effects of feed temperature, heat of reaction as a function of temperature, heat of vaporization, and heat transfer to the jacket. If we consider the specific heat capacity relatively constant and use the multiplicative rule of integration, we can express the differential equation in the generic form of Equation F-2a in terms of temperature to show the process feedback. The relative magnitude of the terms in the denominator of Equation F-6g determines the feedback sign.

$$dQ_o / dt = C_p * \Sigma(F_i * T_i) - C_p * F_o * T_o + (\Delta Q_r / \Delta T_o) * T_o - H_v * F_v - U * A * (T_o - T_j) \qquad \text{(F-6a)}$$

$$dQ_o / dt = d(C_p * M_o * T_o)/dt = C_p * (dM_o / dt) * T_o + C_p * M_o * (dT_o / dt) \qquad \text{(F-6b)}$$

$$F_f = \Sigma F_i \qquad \text{(F-6c)}$$

$$T_f = \Sigma(F_i * T_i) / \Sigma F_i \qquad \text{(F-6d)}$$

$$C_p * (dM_o / dt) * T_o = C_p * (F_f - F_o) * T_o \qquad \text{(F-6e)}$$

$$C_p*M_o*(dT_o/dt) = C_p*F_f*T_f + H_x*R_x - H_v*F_v + U*A*T_j - [C_p*F_f + \Delta Q_r/\Delta T_o + U*A]*T_o \quad \text{(F-6f)}$$

For the manipulation of jacket temperature to control outlet temperature, the main process time constant (τ_p) is (positive feedback if heat of feed and reaction exceeds product of heat transfer coefficient and area):

$$\tau_p = (C_p*M_o)/[C_p*F_f + \Delta Q_r/\Delta T_o + U*A] \quad \text{(F-6g)}$$

For the manipulation of jacket temperature to control outlet temperature, the process gain (K_p) is:

$$K_p = (U*A)/[C_p*F_f + \Delta Q_r/\Delta T_o + U*A] \quad \text{(F-6h)}$$

For the manipulation of jacket temperature to control outlet temperature, the near integrator gain (K_i) is:

$$K_i = (U*A)/(C_p*M_o) \quad \text{(F-6i)}$$

For the manipulation of feed temperature to control outlet temperature, the process gain (K_p) is:

$$K_p = (C_p*F_f)/[C_p*F_f + \Delta Q_r/\Delta T_o + U*A] \quad \text{(F-6j)}$$

For the manipulation of feed flow to control outlet temperature, the process gain (K_p) is:

$$K_p = (C_p*T_f)/[C_p*F_f + \Delta Q_r/\Delta T_o + U*A] \quad \text{(F-6k)}$$

For manipulation of jacket temperature, the additional small secondary process time constant associated with the heat capacity and mass of the jacket wall is:

$$\tau_p = (C_w*M_w)/[U*A] \quad \text{(F-6l)}$$

Any change in the temperature at the heat transfer surfaces or the feed inlet must be dispersed and back mixed into the volume. This process dead time (θ_p) is the turn over time that can be approximated as the liquid inventory divided by the summation of the feed flow rate (F_f), agitator pumping rate (F_a), recirculation flow rate (F_r), and vapor evolution rate or vapor bubble rate (F_v). Since this turn over time is computed in terms of volumetric flow rates, the liquid mass and the mass flow rates are divided by their respective densities as shown in Equation F-6m.

$$\theta_p = (M_o/\rho_o)/[(F_f + F_a + F_r)/\rho_o + F_v/\rho_v] \tag{F-6m}$$

If there is an injector (dip tube or sparger ring) volume, a change in composition at the nozzle must propagate by plug flow to the discharge points of the dip tube or sparger ring. The dead time for a feed flow (F_1) is the injector volume (V_1) divided by the injection mass flow (F_1) divided by its respective density (ρ_1).

$$\theta_p = V_1/(F_1/\rho_1) \tag{F-6n}$$

The ordinary differential equation for the accumulation of liquid reactant mass (M_A) as shown in Equation F-7a includes the effects of feeds (F_i) with a reactant mass fraction (X_{Ai}), reaction rate (R_x), and outlet flow (F_o). The feeds can be from raw material, intermediate products, recycle streams, or multi-stage reactors. If we use the multiplicative rule of integration, we can express the differential equation in the generic form of Equation F-2a in terms of concentration to show the process feedback.

$$dM_A/dt = \Sigma(F_i * X_{Ai}) - (R_x + F_o) * X_{Ao} \tag{F-7a}$$

$$dM_A/dt = d(M_o * X_{Ao})/dt = (dM_o/dt) * X_{Ao} + M_o * (dX_{Ao}/dt) \tag{F-7b}$$

$$F_f = \Sigma F_i \tag{F-7c}$$

$$X_{Af} = \Sigma(F_i * X_{Ai})/\Sigma F_i \tag{F-7d}$$

$$(dM_o/dt) * X_{Ao} = (F_f - F_o) * X_{Ao} \tag{F-7e}$$

$$M_o * (dX_{Ao}/dt) = F_f * X_{Af} - (R_x + F_f) * X_{Ao} \tag{F-7f}$$

For the manipulation of feed flow to control reactant concentration (X_{Ao}), the main process time constant (τ_p) is:

$$\tau_p = M_o/(R_x + F_f) \tag{F-7g}$$

For the manipulation of feed flow to control reactant concentration (X_{Ao}), the process gain (K_p) is:

$$K_p = X_{Af}/(R_x + F_f) \tag{F-7h}$$

For the manipulation of feed flow to control reactant concentration, the near integrator gain (K_i) is:

$$K_i = X_{Af} / M_o \quad \text{(F-7i)}$$

For the manipulation of feed concentration to control reactant concentration, the process gain (K_p) is:

$$K_p = F_f / (R_x + F_f) \quad \text{(F-7j)}$$

The process dead times from turnover time and from feed injection are the same as computed in the section for temperature control (Equations F-6m and F-6n).

Plug Flow Volumes

For plug flow volumes where different streams are being combined, the process gain for controlling the temperature (T_f) or composition (X_{Af}) of the mixture (often a feed to a downstream equipment) can be computed by taking the derivative of Equations F-6d and F-7d with respect to the manipulated flow stream 1 (F_1) to give Equations F-8a and F-8b, respectively. In both cases, the process gain is inversely proportional to total flow (ΣF_i).

$$K_p = dT_f / dF_1 = T_1 / \Sigma F_i \quad \text{(F-8a)}$$

$$K_p = dX_{Af} / dF_1 = X_{A1} / \Sigma F_i \quad \text{(F-8b)}$$

The process dead time for the manipulation of a flow for stream 1 (F_1) is the summation of the injection delay for steam 1 and the piping delay from the point of injection to the point of temperature or composition measurement. For plug flow the residence time, which is the second expression in Equation F-8c completely becomes dead time.

$$\theta_p = V_1 / (F_1 / \rho_1) + V_p / \Sigma(F_i / \rho_i) \quad \text{(F-8c)}$$

The process time constant is essentially zero for true plug flow. For a static mixer there is some back mixing, the residence time in Equation F-8c is split between a dead time and time constant per Equations F-8d and F-8e.

$$\theta_p = V_1 / (F_1 / \rho_1) + x * V_p / \Sigma(F_i / \rho_i) \tag{F-8d}$$

$$\tau_p = (1 - x) * V_p / \Sigma(F_i / \rho_i) \tag{F-8e}$$

It is obvious from the above that both the process gain and dead time are inversely proportional to total flow.

Controller Tuning

The implication of the results can be best seen if Lambda is set equal to the total loop dead time (θ_o) resulting in Equation F-9a for the controller gain. If the open loop time constant (τ_o) is large compared to the dead time, the ratio of the open loop gain (K_o) to the time constant is the "near integrator" gain (K_i) shown in Equation F-9b which is the controller gain touted by the Simplified Internal Model Control (SIMC) as providing the best disturbance rejection. This controller gain is ½ of the gain from the Ziegler Nichols reaction curve method.

$$K_c = 0.5 * \frac{\tau_o}{K_o * \theta_o} \tag{F-9a}$$

$$K_c = 0.5 * \frac{1}{K_i * \theta_o} \tag{F-9b}$$

The time constant (τ_o) in the numerator of Equation F-9a is the largest time constant in the loop wherever it occurs. Hopefully, the process is mixed well enough and the instrumentation is fast enough that the largest time constant is in the process ($\tau_o = \tau_p$) and not the automation system. A large time constant in the process slows down the disturbance and is desirable. A large time constant in the measurement and final element is detrimental because it slows down the ability of the controller to see and react to disturbance, respectively.

The open loop gain (K_o) in the denominator is dimensionless. The process gain is actually the product of the manipulated variable gain, the process gain (K_F), the gain of nonlinear process variables, and the controlled variable gain. For a loop that throttles a control valve, the manipulated variable gain is the slope of the valve's installed characteristic. For the primary loop of a cascade control system, the manipulated variable gain is the secondary loop set point span divided by 100%. The controlled variable gain is 100% divided by the process variable span. Thus, changes in calibration span affect the computed controller gain.

Finally, the dead time (θ_o) in the denominator is really the total loop dead time, which is summation of the process dead time (θ_p) plus all the small time lags and delays in the loop. While the names open loop time constant (τ_o), open loop gain (K_o), and total loop dead time (θ_o) for the parameters in Equation F-10a are more definitive, nearly all of the control literature uses the terms process time constant, process gain, and process gain indiscriminately.

Nomenclature

Process Parameters:

A_o = cross-sectional area of liquid level (m^2)

A = heat transfer surface area (m^2)

C_p = heat capacity of process (kJ/kg*°C)

C_w = heat capacity of wall of heat transfer surface (kJ/kg*°C)

F_a = agitator pumping rate (kg/sec)

F_f = total feed flow (kg/sec)

F_g = gas flow (kg/sec)

F_i = feed stream i flow (kg/sec)

F_o = vessel outlet flow (kg/sec)

F_r = recirculation flow (kg/sec)

F_v = vaporization rate (kg/sec)

H_v = heat of vaporization (kJ/kg)

H_x = heat of reaction (kJ/kg)

L_o = liquid level (m)

M_A = component A mass (kg)

M_g = gas mass (kg)

M_o = liquid mass (kg)

M_w = mass of wall of heat transfer surface (kg)

P_g = gas pressure (kPa)

T_f = total feed temperature (°C)

T_g = gas temperature (°C)

T_i = feed stream i temperature (°C)

T_o = vessel outlet temperature (°C)

t = time (sec)

Q_o = total heat of liquid (kJ)

Q_r = heat from reaction (kJ)

R_x = reaction rate (kg/sec)

R_g = universal constant for ideal gas law (kPa*m³/°C)

ρ_g = gas density (kg/m³)

ρ_i = stream i density (kg/m³)

ρ_o = liquid density (kg/m³)

ρ_v = density of vapor (kg/m³)

U = overall heat transfer coefficient (kJ/m²*°C)

V_g = gas volume (m³)

V_i = injection (e.g. dip tube or sparger ring) volume (m³)

V_p = piping volume (m³)

x = fraction of volume that is plug flow

X_{Af} = total feed component A concentration (mass fraction)

X_{Ai} = feed stream i component A concentration (mass fraction)

X_{Ao} = vessel outlet component A concentration (mass fraction)

Generic Terms:

X = process input (manipulated variable) (eu)

Y = process output (controlled variable) (eu)

Dynamic Parameters:

K_c = PID controller gain (dimensionless)

K_i = integrating process gain (1/sec)

K_o = open loop gain (dimensionless)

K_p = process gain (eu/eu)

τ_p = process time constant (sec)

τ_o = open loop time constant (sec)

θ_p = process dead time (sec)

θ_o = total loop dead time (sec)

References

1. Boudreau, Michael, A., McMillan, Gregory K., and Wilson, Grant E., "Maximizing PAT Benefits from Bioprocess Modeling and Control", Pharmaceutical technology IT Innovations Supplement, November, 2006.

2. Boudreau, Michael, A. and McMillan, Gregory K., *New Directions in Bioprocess Modeling and Control – Maximizing Process Analytical Technology Benefits*, Instrumentation, Automation, and Systems (ISA), 2006.

3. McMillan, Gregory, *Good Tuning – a Pocket Guide*, 2nd edition, Instrumentation, Automation, and Systems (ISA), 2005.

4. McMillan, Gregory and Cameron, Robert, *Models Unleashed – Virtual Plant and Model Predictive Control Applications*, Instrumentation, Automation, and Systems (ISA), 2004.

5. Blevins, Terrence L., McMillan, Gregory K., Wojsznis, Willy K., and Brown, Michael W., *Advanced Control Unleashed – Plant Performance Management for Optimum Benefits*, Instrumentation, Automation, and Systems (ISA), 2003.

Appendix G
FORTRAN Subroutine for Dynamic Simulation of Extruders

```
      SUBROUTINE ZEXTRD (Fo,Po,To,Kp,Ko,M,J,E,T,Q,Z,L,R,H,
      S,X,Ri,Ro, $Wc,Wf,Ks,Kd,Pi,Ti,Ci,Di,Px,Tx,Co,Do,Vr,Tr,Kr,Kc,
      delt,time,N)
C
      implicit none
      integer*4 I,N
      real*4 M(20),J(20),E(20),T(20),Z(20),L(20),R(20),H(20)
      real*4 C(20),D(20),V(20)
      real*4 Fo,Po,To,Kp,Ko,S,X,Ri,Ro,Wc,Wf,Ks,Kd,Pi,Ti,Ci,Di
      real*4 Px,Tx,Co,Do,Vr,Tr,Kr,Kc
      real*4 delt,time
      real*4 Vm,Lx,Lz,Hm,Rm,Pm,Qs,Qf,MC,UA,TC
      real*4 A,B
      real*4 dQdt(20),Q(20)
C
C     This subroutine gives outlet flow, pressure, and temperature
C     and zone temperatures in an extruder using equations from
C     pages 35-36 in CEP, Nov 1988 and 618-622 in Polymer, June 1977
C
C     User must initialize Fo, Po, and To for hot or warm start-up!
C     (Fo > 0.0 and Po > Px and To >Ti)
C     M, J, E, T, Q wlll be initialized by subroutine!
C
C     Variables Computed by Subroutine
C     Fo is polymer volumetric flow at extruder outlet (ft^3/min)
C     Po is polymer pressure at extruder outlet (lb/in.^2)
```

```
C     To is polymer temperature at extruder outlet (lb/in.^2)
C     Kp is correction factor for pressure calculation (0.1 - > 10.)
C     Ko is correction factor for flow calculation (0.1 - > 10.)
C     M is melt sensible heat input in zone i (Btu/min)
C     J is jacket thermal lagged heat input in zone i (Btu/min)
C     E is energy generation from speed in zone i (Btu/min)
C     T is polymer temperature in zone i (°F)
C     Q is heat accumulation in zone i (Btu)
C
C     Variable Values Supplied by User:
C     Z is net heat from barrel and screw in zone i (Btu/min)
C     L is length of zone i (inches)
C     R is radius of screw in zone i (inches)
C     H is channel height in zone i (inches)
C     S is speed of screw (rpm)
C     X is helix angle (radians)
C     Ri is inside radius of barrel (inches)
C     Ro is outside radius of barrel (inches)
C     Wc is width of channel (inches)
C     Wf is width of flight (inches)
C     Ks is factor for viscous energy generation (0.1 - > 10.)
C     Kd is factor for die (in.^3)
C     Pi is pressure at inlet (Ib/in.^2)
C     Ti is temperature at inlet (°F)
C     Ci is heat capacity of polymer at inlet (Btu/lb*°F)
C     Di is density of polymer at inlet (lb/ft^3)
C     Px is pressure at die exit (lb/in.^2)
C     Tx is maximum possible temperature (°F)
C     Co is heat capacity of polymer at outlet (Btu/lb b*°F)
C     Do is density of polymer at outlet (lb/ft^3)
C     Vr is viscosity at reference temperature (cp)
C     Tr is reference temperature (°F)
C     Kr is exponential factor for viscosity function (0.1 - > 10.)
C     Kc is filter factor to break Fo and Po cycles (about 0.05)
C     delt is integration step size (minutes)
C     time is program run time (minutes)
C     N is number of zones (zones may be combined)
C
C     Variables Computed for Internal Use:
C     V is absolute viscosity in zone i (lbf*min/in.^2)
```

```
C     Vm is absolute viscosity mean value (lbf*min/in.2)
C     Lx is total length of zones (inches)
C     Lz is progressive length of zone (inches)
C     C is heat capacity of polymer in zone i (Btu/lb*°F)
C     D is density of polymer in zone i (lb/ft3)
C     Hm is channel height mean (inches)
C     Rm is screw radius mean (inches)
C     Pm is pressure rise for mean values (lb/in.2)
C     MC is product of metal mass and heat capacity (Btu/°F)
C     UA is product of heat transfer coeff and area (Btu/°F*min)
C     TC is thermal time constant of metal mass (min)
C     Xz is number of flights in zone
C     Qs is energy generation from screw channels (ft*lbf/sec)
C     Qf is energy generation from flight clearance (ft*lbf/sec)
C     A is cross-sectional area for polymer flow (ft3)
C     B is base for temperature from heat accumulation (°F/Btu)
C     dQdt is heat derivative in zone i (Btu/min)
C
C     Compute total length of zones:
C
      I = 1
      Lx = 0.0
      do while (I .le. N)
      Lx = Lx + L(I)
      I = I + 1
      end do
C
C     Compute heat capacity, density, and initial temperature in each
      zone
C
      I = 1
      Lz = 0.0
      do while (I .le. N)
      Lz = Lz + L(I)
      C(I) = (Co – Ci)*(Lz/Lx) + Ci
      D(I) = (Do – Di)*(Lz/Lx) + Di
      if (time .lt. delt) then
      T(I) = (To – Ti)*(Lz/Lx) + Ti
      else
      endif
```

```
      I = I + 1
      end do
C
C     Compute absolute viscosity in each zone and mean value
C     (viscosity conversion: lbf*sec/sqin = centipoise/6894760.)
C     (viscosity conversion: lbf*min/sqin = lbf*sec/in.^2/ 60.)
C
      I = 1
      do while (I .le. N)
      V(I) = Vr*exp(-0.01*Kr*(T(I) – Tr))/(6894760.*60)
      I = I + 1
      end do
      Vm = ((V(N) – V(l))/2. + V(l))
C
C     Compute mean pressure
C     (flow conversion: cuin/min = 1728.*ft^3/min)
C
      Hm = (H(N) – H(1))/2. + H(l)
      Rm = (R(N) – R(1))/2. + R(l)
      if (Po .lt. 1.1*Px) Po = 1.1*Px
      if (time .lt. delt) then
      Kp = Po/(12.*Vm*(Wc*Hm*3.14*Rm*S*cos(X) – 1728.*Fo)/
      & (Rm*Wc*Hm))
      else
      Pm = Kp*(12.*Vm*(Wc*Hm*3.14*Rm*S*cos(X) – 1728.*Fo)/
      & (Rm*Wc*Hm))
      endif
C
C     Compute volumetric outlet pressure and flow
C     (conversion: ft^3/min = in.^3/min/1728.)
C
      if (time .lt. delt) then
      Ko = Fo/(Kd/1728.*(Po – Px)/V(N))
      else
      Po = Po + Kc*(Pm + Pi – Po)
      endif
      Fo = Ko*(Kd/1728.*(Po – Px)/V(N))
C
C     Compute thermal lag from metal mass in each zone
C     (length cancels out and U is proportional to S)
```

```
C      (conversion ft^2 =in.^2/144.)
C      (conversion ft = inch/12.)
C
       MC = 488.*(3.14*(Ro**2 – Ri**2)/144.)*0.11
       UA = S*2.*(3.14*Ri)/12.
       TC = MC/UA
       if (TC .lt. 10.*delt) TC = 10.*delt
C
C      Compute energy balance derivative
C      (density conversion: ft^3/min = 1728.*in.^3/min)
C      (speed conversion: rev/sec = rev/min/60.)
C      (energy conversion ft*lbf/min = in*lbf/min/12.)
C      (heat conversion Btu/min = ft*lbf/min/777.65)
C      (L(I)/Wc is approx number of flights in zone i)
C
       I = 1
       do while (I .le. N)
       Qs = V(I)*(((3.14*2.*R(I)*S)**2)*(Wc*L(I))/(H(I)*sin(X)))*Ks
       Qf = V(I)*(((3.14*2.*R(I)*S)**2)/(Ri-R(I)))*Wf*(L(I)*sin(X))
       if (time .lt. delt) then
       J(I) = Z(I)
       else
       J(I) = J(I) + (delt/(delt + TC))*(Z(I) – J(I))
       endif
       E(I) = (L(I)/Wc)*(Qs + Qf)/(777.65*12.)
       if (I .lt. 2) then
       M(I) = 1728.*(Fo*Di*Ci*Ti – Fo*D(I)*C(I)*T(I))
       else
       M(I) =1728.*(Fo*D(I – 1)*C(I – 1)*T(I – 1) – Fo*D(I)*C(I)*T(I))
       endif
       dQdt(I) = J(I) + E(I) + M(I)
       I = I + 1
       end do
C
C      Integrate to get temperature in each zone
       (conversion ft^3 = in.^3/144.)
C      (conversion ft = inch/12.)
C
       I = 1
       do while (I .le. N)
```

```
A = 3.14*(Ri**2 – R(I)**2)/144.
B = A*(L(I)/12.)*D(I)*C(I)
if (time .lt. delt) then
Q(I) = T(I)*B
else
Q(I) = dQdt(I)*delt + Q(I)
endif
if (Q(I) .lt. Ti*B) Q(I) = Ti*B
if (Q(I) .gt. Tx*B) Q(I) = Tx*B
T(I) = Q(I)/B
I = I + 1
end do
To = T(N)
return
end
```

Appendix H
Convective Heat Transfer Coefficients

1. For heating and cooling liquids flowing perpendicular to a single cylinder (e.g., thermowell), the convective heat transfer coefficient can be estimated as follows (note that the viscosity and velocity use hours for time units) [1]:

$$h = [k/(2 * r)] * \{0.35 + 0.56 * [(2 * r * v * d)/u]^{0.52}\} * [(c * u)/k]^{0.3} \quad \text{(H-1)}$$

where:

c = heat capacity of fluid (Btu/lb * °F)

d = density of fluid (lb/ft^3)

r = outside radius of thermowell (ft)

k = thermal conductivity of fluid (Btu/hr * ft * °F)

u = viscosity of fluid (lb/ft * hr)

v = velocity of fluid (ft/hr)

2. For the more general thermowell case that covers a wide range of Reynolds numbers and fluids, the convective heat transfer coefficient can be estimated as follows (note that viscosity is in centipoise and velocity is in ft/sec) [2]:

$$h = 3960 * v * d * c * Cj * Re^{(Nj - 1)} * Pr^{(-0.69)} \quad \text{(H-2)}$$

$$Re = (248 * r * v * d)/u \quad \text{(H-3)}$$

$$Pr = (2.42 * u * c)/k \text{ (liquids)} \quad \text{(H-4)}$$

$$Pr = 0.735 \text{ (gases)} \quad \text{(H-5)}$$

where:

	Cj	Nj
Re < 4	0.891	0.330
4 < Re < 40	0.821	0.385
40 < Re < 4,000	0.615	0.466
4,000 < Re < 40,000	0.174	0.618
40,000 < Re	0.024	0.805

c = heat capacity of fluid (Btu/lb * °F)

d = density of fluid (lb/ft^3)

r = outside radius of thermowell (ft)

k = thermal conductivity of fluid (Btu/hr * ft * °F)

Pr = Prandtl number (dimensionless)

Re = Reynolds number (dimensionless)

u = viscosity of fluid (cp)

v = velocity of fluid (ft/sec)

3. For the flow of fluids in tubes where the Reynolds number is between 10,000 and 120,000, the convective heat transfer coefficient can be estimated as follows (note that viscosity and velocity use seconds for time units) [3]:

$$h = 0.023 * v^{0.8} * (2 * r)^{-0.2} * k * (u/d)^{-0.8} \quad \text{(H-6)}$$

where:

d = density of fluid (lb/ft^3)

r = inside radius of tube (ft)

k = thermal conductivity of fluid (Btu/hr * ft * °F)

u = viscosity of fluid (lb/ft * sec)

v = velocity of fluid (ft/sec)

Table H-1. The convective heat transfer coefficient range changes with fluid type, turbulence, and phase [3]

Fluid and Condition	h Range[2] (Btu/hr * ft^2 * °F)
Air-free convection	1 - 5
Air- or superheated steam-forced convection	5 - 50
Oils-forced convection	10 - 30
Water-forced convection	50 - 2000
Water boiling	500 - 10000
Steam condensing	1000-20000

References

1. Kardos, P. W., "Response of Temperature Measuring Elements," *Chemical Engineering*, August 29, 1977, pg. 82.

2. Crawford, C. L., "Thermowell heat Conduction Error Versus," Texas A&M Symposium, January 1982, Appendix I, pg. 2.

3. Kreith, F., *Principles of Heat Transfer*, 2nd edition, International Textbook Company, 1965, pg. 376.

Appendix I
Implementation Checklist for Best Performance

In the following list, increases in on-stream time can increase efficiency besides capacity by eliminating the time and off-spec and waste associated with abnormal operations, startup, and shutdown. An increase in yield or decrease in recycle can be taken as a decrease in raw material costs (same production rate for lower feed rate) or an increase in production rate (higher production rate for the same feed rate). The order of the list is in order of things to check and somewhat in the order of priorities.

1. **Use smart transmitters with the best sensor technology and lead length and extension wire compensation (minimize system measurement error)**

    a. Seek sensor and transmitter with the best sensitivity and repeatability

    b. Use high thermal conductivity materials between sensor and process (avoid ceramic protection tubes and glass lined baffles)

    c. Maximize thermowell insertion length and minimize air gap between sensor and inside wall of thermowell

    d. Use locally mounted or integral mounted transmitters

2. **Pick sensor location and installation method to provide the most representative measurement with no stagnation, best velocity, fastest response, and least noise (minimize location errors)**

    a. Increase velocity to reduce response time and coatings but not so high to cause abrasion or vibration

    b. Pick location with good mixing, minimal transportation delay, and least bubbles, slime, and solids

    c. Use wireless transmitters to facilitate optimization of location

3. **Use real throttle valves with smart positioners (minimize limit cycles)**

    a. Avoid on-off and isolation valves posing as throttling valves. Go to a control valve manufacturer instead of a piping valve manufacturer

    b. Seek actuator, positioner, and valve type with best sensitivity of installed flow characteristic and signal response with least stick-slip and backlash

    c. Verify positioner feedback measurement is representative of internal closure member (e.g., ball, disk, or plug) and not just actuator position

4. **Tune control loop with on-demand auto tuner or adaptive controller to loop objectives for performance objectives. Tuning speed is chosen to:**

    a. Ensure an exceptionally smooth PV and output response by decreasing transfer of variability from PV to output (increasing controller gain) for:

        - level loops on surge tanks to minimize feed upsets
        - dead time dominant loops (dead time >> process time constant)
        - interacting loops (e.g., headers)

- loops on piping or equipment with no back mixing (e.g., blenders, heat exchangers, extruders, static mixers, sheets, webs, and yarns)

b. Provide good load rejection of moderately fast disturbances by increasing transfer of variability from PV to output (increasing controller gain) for:

- Fed-batch and continuous agitated vessel and column loops

c. Provide good load rejection of extremely fast disturbances by setting the gain and reset as a factor of dead time rather than the time constant for:

- Continuous agitated vessel and column loops

d. Provide minimal overshoot of set points and off-spec in loops dominated by a slow lag (process time constant >> loop dead time and slower than 10 minutes) by tuning the loops as near-integrating processes for:

- Fed-batch and continuous agitated vessels and column loops (set point changes occur at startup or for changes in batch phase and product grade)

e. Provide minimal peak error by maximizing controller gain even if it requires increasing reset time to increase onstream time by:

- Prevention of Safety Instrument System (SIS) activation
- Prevention of environmental violation
- Prevention of equipment damage

5. **Add DCS signal filter or damping adjustment to keep loop output fluctuations from noise less than the valve dead band to prevent excessive valve packing wear and inflicting disturbances on loop. For wireless transmitters use damping adjustment to reduce keep transmitter output fluctuations from noise less than wireless trigger level for exception reporting to eliminate unnecessary communication and extend battery life.**

6. **Eliminate on-off actions to eliminate fast large disturbances**

    a. Replace on-off control by switches with loops

    b. Eliminate manual actions by adding loops, keeping loops in highest design mode, adding feedforward, and automating and tuning loops to handle startup and abnormal operating conditions

    c. Replace pure batch with fed-batch automation by replacing discrete sequential actions (e.g., stepping feeds) with loops (e.g., throttling feeds)

7. **Tune loops that create feed disturbances (e.g., surge level loops) to provide a smooth slow transition in feed rate**

8. **Add cascade control to compensate for nonlinearities and pressure disturbances** (e.g., secondary flow loop and secondary coolant temperature loop)

9. **Add feedforward control of measurable fast disturbances not compensated by secondary loop**

10. **Optimize set points by operating closer to constraints for production rate or product quality spec.**

    a. Eliminate operating margin imposed by shift's perceived sweet spot or operating margin caused by process variability from not doing check list items 1-9

    b. Find more efficient operating points based on R&D reports and virtual plant exploration - confirm with process tests

    c. Add model predictive control to optimize set points as process conditions and market requirements change

Index

Printed and bound by CPI Group (UK) Ltd, Croydon, CR0 4YY

13/09/2026

14961057-0002